Engineering Technologies

Engineering Technologies covers the mandatory units for the EAL Level 3 Diploma in Engineering and Technology:

- Each compulsory unit is covered in detail with activities, case studies and self-test questions where relevant.
- Review questions are provided at the end of each chapter and a sample multiple-choice examination is included at the end of the book.
- The book has been written to ensure it covers what learners need to know.
- Answers to selected questions in the book, together with a wealth of supporting resources, can be found on the book's companion website. Numerical answers are provided in the book itself.

Written specifically for the EAL Level 3 Diploma in Engineering and Technology, this book covers the two mandatory units: Engineering and Environmental Health and Safety, and Engineering Organizational Efficiency and Improvement. Within each unit, the learning outcomes are covered in detail and the book includes activities and 'Test your knowledge' sections to check your understanding. At the end of each chapter is a checklist to make sure you have achieved each objective before you move on to the next section. At www.key2engtech.com, you can download answers to selected questions found within the book, as well as reference material and resources.

This book is a 'must-have' for all learners studying for their EAL Level 3 Diploma award in Engineering and Technology.

Mike Tooley has over 30 years' experience of teaching engineering, electronics and avionics to engineers and technicians, previously as Head of the Department of Engineering, Faculty of Technology and Vice Principal at Brooklands College. Mike currently works as a consultant and freelance technical author.

Engineering
Technologies   Level 3

Mike Tooley

Routledge
Taylor & Francis Group

LONDON AND NEW YORK

First published 2017 by Routledge

2 Park Square, Milton Park, Abingdon, Oxfordshire OX14 4RN

52 Vanderbilt Avenue, New York, NY 10017

Routledge is an imprint of the Taylor & Francis Group, an informa business

First issued in paperback 2019

British Library Cataloguing in Publication Data
A catalogue record for this book is available from the British Library

Library of Congress Cataloging in Publication Data
Names: Tooley, Michael H., author.
Title: Engineering technologies. Level 3 / Mike Tooley.
Description: Boca Raton : CRC Press, Taylor & Francis, 2017. | Includes index.
Identifiers: LCCN 2016040330| ISBN 9781138674929 (paperback) | ISBN 9781315560960 (ebook)
Subjects: LCSH: Engineering.
Classification: LCC TA147 .T6696 2017 | DDC 620--dc23
LC record available at https://lccn.loc.gov/2016040330

ISBN: 978-1-138-67492-9 (pbk)

Typeset in Helvetica by
Servis Filmsetting Ltd, Stockport, Cheshire
Printed and bound by CPI Group (UK) Ltd, Croydon CR0 4YY

Contents

Preface

Welcome to the challenging and exciting world of engineering! This book is designed to help you succeed on a course leading to the EAL Level 3 Diploma in Engineering and Technology. It contains all of the essential underpinning knowledge required of a student who may never have studied engineering before or who wishes to build on an existing Level 2 qualification such as the EAL Level 2 Diploma in Engineering and Technology.

About you

Have you got what it takes to be an engineer? The EAL Level 3 Diploma in Engineering Technology will help you find out and still keep your options open. The Diploma is often used as the technical element of the foundation modern apprenticeship framework. Successful completion of the course will provide you with a route into studying engineering at Level 4, for example, a Foundation Degree in Engineering or a BTEC Higher National award in Engineering.

Engineering is an immensely diverse field but, to put it simply, engineering, in whatever area that you choose, is about thinking *and* doing. The 'thinking' that an engineer does is both logical and systematic. The 'doing' that an engineer does can be anything from building a bridge to testing a space vehicle. In either case, the essential engineering skills are the same. You do not need to have studied engineering before starting a Level 3 programme. All that is required to successfully complete the course is an enquiring mind, an interest in engineering, and the ability to explore new ideas in a systematic way. You also need to be able to express your ideas and communicate these in a clear and logical way to other people.

As you study your EAL Level 3 course in Engineering and Technology you will be learning in a practical environment as well as in a classroom. This will help you to put into practice the things that you learn in a formal class situation. You will also discover that engineering is fun – it's not just about learning a whole lot of meaningless facts and figures!

How to use this book

This book provides full coverage of the two mandatory units of the EAL Level 3 Diploma in Engineering and Technology. The two mandatory units are entitled Engineering and environmental health and safety, and Engineering organizational efficiency and improvement. Within the book, each chapter is devoted to a major 'sub-group' topic. The book includes text, illustrations, test your knowledge questions, examples and activities (where relevant). Each chapter concludes with a set of review questions. Answers to selected review questions can be downloaded from the author's website, www.key2engtech.com.

'Test your knowledge' questions are interspersed with the text throughout the book. These questions allow you to check your understanding of the preceding text. They also provide you with an opportunity to reflect on what you have learned and consolidate this in manageable chunks.

Most 'Test your knowledge' questions can be answered in only a few minutes and the necessary information can be gleaned from the surrounding text. Activities, on the other hand, require a significantly greater amount of time to complete and they are designed to be completed outside the classroom, often in a workshop or other practical environment. They may also require additional library or resource area research coupled with access to computing and other ICT resources. Don't expect to complete *all* of the activities in this book – your teacher or lecturer will ensure that those activities you do undertake relate to the resources available to you and that they can be completed within the timescale of the course. Activities make excellent vehicles for improving your skills and for gathering the evidence that can be used to demonstrate that you are competent in a range of core engineering skills.

To bring some difficult topics into sharp focus we've included a number of Case studies. These deal with issues such as health and safety policy, risk assessment, permits to work, production planning, statistical process control, and continuous improvement. The activities and questions that follow the case studies are designed to aid your understanding and give you plenty to think about.

The Review questions presented at the end of each chapter are designed to provide you with an opportunity to test your understanding of each unit. These questions can be used for revision or as a means of generating a checklist of topics with which you need to be familiar. Here again, your tutor may suggest that you answer specific questions that relate to the context in which you are studying the course.

The book ends with some useful information and data presented in the form of four appendices. These include sample assessment questions, abbreviations for common terms used in engineering, information on how to use a scientific calculator, and a list of useful websites. To provide you with an indication of the standard that you need to reach, Appendix 1 provides you with a representative set of questions. More resources are available at the author's website, www.key2engtech.com.

Finally, here are a few general points worth keeping in mind:

- Allow regular time for reading – get into the habit of setting aside an hour or two at the weekend to take a second look at the topics that you have covered during the week.
- Make notes and file these away neatly for future reference – lists of facts, definitions and formulae are particularly useful for revision!
- Look out for the interrelationship between units and topics – you will find many ideas and a number of themes that crop up in different places and in different units. These can often help to reinforce your understanding.
- Don't be afraid to put your new ideas into practice. Remember that engineering is about thinking *and* doing – so get out there and *do* it!

Good luck with your EAL Level 3 Diploma in Engineering and Technology!

Mike Tooley

UNIT **1**

Engineering and environmental health and safety

CHAPTER 1

Health and safety: roles and responsibilities

Learning outcomes

When you have completed this chapter you should understand the requirements of an engineering organization in meeting health and safety legislation and regulations, including being able to:

1.1 Recognize the roles of key people involved in workplace health and safety.

1.2 Recognize the roles of organizations involved in workplace health and safety.

1.3 State the key duties of the employee in conforming with health and safety requirements.

1.4 State the key duties of an employer in the management of health and safety.

1.5 Recognize the content and application of key health and safety legislation.

Chapter summary

The ability to work safely in an engineering environment is essential not only for your own comfort and safety but also for the safety of those around you. This chapter will introduce you to the legislation and safety regulations that govern working practices in engineering. As you work through the chapter you will find a number of hands-on activities that will help you appreciate some of the potential hazards that exist in the workplace and the ways in which they can be minimized. Importantly, each of these activities will take you out of the classroom, preparing you for work in a real engineering environment.

Learning outcome 1.1

Recognize the roles of key people involved in workplace health and safety

Everyone has a role to play in ensuring that the workplace is a safe and healthy place. However, some people have specialist roles and key responsibilities that you need to be aware of. Of particular importance are:

- Inspectors appointed by the Health and Safety Executive (HSE)
- Designated Safety Officers
- Health and Safety Representatives
- Environmental Health Officers.

All of these people have key roles in ensuring that the working environment is safe and they all have specific responsibilities coupled with knowledge and experience to make them effective in their roles. We will briefly look at each of the roles.

Health and Safety Executive Inspectors

Health and Safety Executive (HSE) Inspectors conduct visits to engineering companies (and other businesses) to ensure that they comply with all aspects of health and safety laws and that workplaces are not the cause of ill health, injury or even death. HSE Inspectors will advise employers and investigate accidents, ensuring that the law is applied and, where necessary, enforced.

In particular, HSE Inspectors will be looking for evidence that a company has carried out an assessment of the risks associated with its activities. We will be looking at risk assessment later on but for now you just need to be aware that the key elements associated with an engineering activity or process are to ensure that:

- all hazards and risks are clearly identified
- an evaluation is carried out and any gaps are identified
- relevant workers are involved in the process of identification and evaluation
- effective solutions are developed in order to avoid hazards and minimize risks.

Designated Safety Officers

Designated Safety Officers liaise with employers, employees, directors and trade unions using their skills, knowledge and experience to promote a positive health and safety culture in the workplace. Because of their unique position they have a first-line role in ensuring that employers and workers comply with relevant safety legislation and that safety policies and practices are effective within the organization. Their roles and responsibilities may vary depending on the type and scope of the organization but in an engineering company they often include:

- carrying out risk assessments and considering how risks could be reduced
- outlining safe operational procedures which identify and take account of all relevant hazards
- carrying out regular site inspections, checking that policies and procedures are being properly implemented
- making changes to working practices, ensuring that they are safe and comply with relevant legislation
- preparing health and safety strategies and developing internal policy
- leading in-house training with managers and employees about health and safety issues and risks
- keeping records of inspection findings and producing reports (including suggestions for improvement)
- recording of incidents and accidents and producing statistics for managers
- keeping up to date with new legislation and maintaining a working knowledge of all Health and Safety Executive (HSE) legislation and any developments that affect the employer's industry
- attending Institution of Occupational Safety and Health (IOSH) seminars and reading professional journals
- producing management reports, newsletters and bulletins
- ensuring the safe installation of equipment
- managing and organizing the safe disposal of hazardous substances, e.g. asbestos

Key point

HSE Inspectors conduct visits to engineering companies (and other businesses) to ensure that they comply with all aspects of relevant health and safety legislation.

Key point

Safety Officers promote a positive health and safety culture in the workplace, carrying out risk assessments and ensuring that operational practices and procedures take into account potential hazards.

- advising on a range of specialist areas, e.g. fire regulations, hazardous substances, noise, safeguarding machinery and occupational diseases.

Health and Safety Representatives

Depending upon the size and scope of an organization, one or more Health and Safety Representatives also have a key role to play in workplace health and safety. Larger organizations may have a Safety Committee which will normally include elected Health and Safety Representatives together with a Health and Safety Officer and a senior management representative, such as a Production Manager or Engineering Manager.

Health and Safety Representatives are usually elected by their colleagues or they may be appointed by a trade union and, as such, they represent the interests of those who are actually performing engineering operations. This is a very different role from the employer-appointed Health and Safety Officer who will have health and safety duties as part of or all of his or her job description.

Health and Safety Representatives usually have the following responsibilities:

- advising and communicating health and safety policy within the organization
- encouraging worker participation in training and development activities related to health and safety
- raising concerns and complaints from workers related to health and safety issues
- assisting with workplace inspections
- informing colleagues on matters relevant to health, safety and welfare.

Being a Health and Safety Representative is something that you might want to consider for your own personal development. The role will help you understand the importance of having an effective health and safety policy and you will learn a great deal from it. So, if the opportunity presents itself, why not put yourself forward?

Environmental Health Officers

Environmental Health Officers (sometimes also known as Public Health Inspectors) are responsible for investigating incidents that affect health such as pollution, accidents at work, noise control, toxic contamination, pest infestations, food poisoning, and waste management. Their role involves advisory work, education and law enforcement. In an engineering context their responsibilities include:

Key point

Health and Safety Representatives are elected by their colleagues or appointed by trade unions and they represent the interests of those who are actually performing engineering operations. They encourage worker participation in health and safety activities and, where appropriate, they can raise concerns and complaints.

- inspections and compiling reports of their findings
- providing advice and training
- gathering samples to be tested
- investigating complaints
- serving legal notices
- providing evidence in court
- liaising with other organizations, as appropriate.

Environmental Health Officers (EHO) have a particular role in ensuring that the workplace and its surroundings are safe, healthy and hygienic. Working in partnership with relevant Government Ministries (such as Health, Agriculture and Environment) as well as local councils, businesses and community groups, they have a major role in protecting public health. This responsibility is particularly important where engineering activities might involve hazardous and toxic waste products, noise and atmospheric pollutants.

Figure 1.1 Various hazards are present in any engineering environment.

> **Key point**
>
> Environmental Health Officers ensure that the workplace and its surroundings are safe, healthy and hygienic. Working in partnership with external bodies such as local councils and community groups they have a major role in protecting public health.

Test your knowledge 1.1

List **three** key responsibilities associated with each of the following health and safety roles:

a) A HSE Inspector
b) A designated Safety Officer
c) A Health and Safety Representative
d) An Environmental Health Officer

Test your knowledge 1.2

In a large engineering company, explain **two** differences between the roles and responsibilities of a Safety Officer and a Health and Safety Representative.

Test your knowledge 1.3

List **three** different job roles and responsibilities that you would expect to be represented within the Safety Committee of a large engineering company. Explain your answer.

Learning outcome 1.2

Recognize the roles of organizations involved in workplace health and safety

There are a number of key organizations concerned with health and safety. The most noteworthy are:

- Health and Safety Commission (HSC)
- Health and Safety Executive (HSE)
- Local authorities
- Trading Standards
- Environmental Health.

The Health and Safety Commission

The Health and Safety at Work Act 1974 (HASAWA) established two key organizations: the Health and Safety Commission (HSC) and the Health and Safety Executive (HSE). The HSC is responsible for making arrangements to secure the health, safety and welfare of people at work and for protecting the general public against risks to their health and safety from work activities. The HSC has a number of duties that are enforced by law. These *statutory duties* include:

- encouraging others to secure safe and healthy working conditions
- arranging for research and training to be carried out
- encouraging others to undertake research and training
- arranging for an advisory and information service for stakeholders
- submitting proposals to government ministers concerning regulations and standards.

You need to be aware that the HSC has wide powers and they include anything necessary to help them perform their statutory duties. The primary role of the HSC is to give strategic direction to the work of the HSE. It also has powers to:

- approve and issue Approved Codes of Practice (ACOP) with the consent of the Secretary of State, subject to consultation with Government departments and other organizations
- make agreements with any Government department or person to perform HSC or HSE functions on HSC/E's behalf
- make agreements with any minister, Government department, or public authority for HSC to perform functions on their behalf
- give mandatory guidance to local authorities on enforcement
- direct HSE or authorize any other person to investigate and report on accidents or other matters, and subject to regulations made by the relevant minister, direct inquiries to be held.

> **Key point**
>
> The Health and Safety Commission (HSC) is responsible for making arrangements to secure the health, safety and welfare of people at work and for protecting the general public against risks to their health and safety from work activities.

The Health and Safety Executive

The Health and Safety Executive (HSE) advises and assists the HSC and, together with local authorities, these key organizations have day-to-day responsibility for:

- enforcing health and safety law
- investigating accidents
- licensing and approving standards
- providing information and guidance
- proposing new laws and standards
- commissioning research.

> **Key point**
>
> The HSC provides strategic direction for the HSE.

Test your knowledge 1.4

List **three** statutory duties of the HSC.

Activity 1.1

Figure 1.2 shows an extract from a recent HSE Safety Alert. Read the bulletin carefully and answer the following questions:

1. What type of component has been found to be faulty?
2. What were the consequences of the failure?
3. What caused the failure?
4. What could have been done to avoid the failure?
5. What should duty holders do to detect faulty components?

Catastrophic failure of a pipework clamp connector

Health and Safety Executive - Safety Alert	
Department Name:	Energy Division (Offshore)
Bulletin No:	ED 2-2015
Issue Date:	18 September 2015
Target Audience:	Offshore Chemical Processing and production
Key Issues:	This Safety Alert highlights the issues of inadequate material properties of small diameter clamp connectors and the potential for sudden brittle failure.

Introduction and background:

An incident occurred on an offshore production platform in December 2014 where a 1" diameter Vector 'Techlok' pipework clamp connector catastrophically failed causing a gas release. The failure was caused by poor heat treatment during the manufacturing process, which led to failure by cracking as shown:

Further enquiry has revealed similar failures on other small-sized (1", 1½" and 2" diameter) clamp connector segments.

Cracking caused by poor heat treatment during the manufacturing process

Reasons for failure:

The primary cause of failure of the clamp was due to hydrogen cracking. Evidence from this clamp and others checked later indicates that high hardness, with values in excess of 48-50HRC (Rockwell hardness value), being the underlying cause. High hardness increases the material's susceptibility to cracking in general from reduced ductility.

Investigations revealed that prior to 2010 the clamps, which require quench and temper heat treatment, were not subject to 100% hardness testing. Hence the quality controls were not sufficient to detect components that had received improper heat treatment and lacked ductility.

Clamp connector segments are produced by a forging process in a number of foundries in the UK and are manufactured from AISI 4140 alloy steel with a recommended Rockwell hardness value of 22HRC.

To date all defective clamps found were produced by one manufacturer (George Dykes) and supplied to Vector Technology Group (Techlok) up until 2010. These were limited to 2" diameter clamps and below. It has subsequently been established that the similar 'G' clamp supplied by Destec Engineering, was also manufactured by George Dykes. A Destec 'G' clamp is also known to have failed in service in 2014.

The confirmed numbers of failed clamps is low and they appear to be random and not batch related, therefore it is not possible to identify them through heat numbers or heat treatment records.

Action required:

Duty Holders should identify if they have in use any 2" or below Techlok or Destec 'G' clamp connectors supplied before 2010, or have the potential to be used (for example spares kept in stores) on any of their installations and to verify their fitness for service.

Verification could be established from supplier records whereby the material properties of the clamps are fully certified, or by appropriate NDT inspection and hardness tests.

Figure 1.2 Extract from an HSE Bulletin. (This image contains public sector information published by the Health and Safety Executive and licensed under the Open Government License.)

Local authorities

In conjunction with the HSE, local authorities are responsible for enforcing health and safety legislation. Together they ensure that organizations manage the health and safety of their workforce and those affected by their work. UK health and safety law places duties on a variety of people. Primary responsibilities are placed on those who create and/or have the greatest control of the risks associated with a particular activity. Those who create the risks at the workplace are responsible for controlling them. The Health and Safety at Work Act 1974 and its associated regulations mainly place duties on employers. Employees also have duties under the Act. In fact, everyone has a part to play in ensuring healthy and safe conditions at work.

Health and safety legislation in Great Britain is enforced by the HSE or one of over 380 local authorities, depending on the main activity carried out at any particular premises. Each local authority is an enforcing authority in its own right and must make adequate provision for enforcement. Local authorities use a number of intervention approaches to regulate and influence businesses in the management of health and safety risks including:

- provision of advice and guidance to individual businesses or groups
- proactive interventions including inspection
- reactive interventions (e.g. to investigate an accident or complaint).

Local authority inspectors may use enforcement powers, including formal *enforcement notices*, to address occupational health and safety risks and secure compliance with the law. Prosecution action may be appropriate to bring duty holders to account for failures to safeguard health and safety.

Trading Standards

Under the Sale of Goods Act 1979, any product must be 'fit for purpose', being of satisfactory quality, and fit its description. This means that a manufacturer's products must fulfil the purpose the customer has been led to expect and the reasons that led them to buy it. The Act also covers any purpose that a customer asks about when the product is purchased and is guaranteed by the retailer to meet that purpose when it is sold. If a product is not fit for purpose, the customer is within their rights to have the goods replaced or repaired. Manufacturers can also be held liable in any legal action for harm caused to consumers or businesses as a result of unintended side-effects or the failure of products manufactured or supplied. In addition, manufacturing and processing systems must comply with relevant environmental laws.

Trading Standards Officers (TSO) are employed by local authorities to advise on consumer law and to investigate complaints raised by customers. They will usually attempt to resolve problems before they escalate but they have powers to prosecute offenders where necessary. Services offered by Trading Standards tend to vary across the UK and complaints are often referred to them by Citizens Advice.

CE marking

A CE mark is a manufacturer's claim that its product meets specified essential safety requirements set out in relevant European directives. Certain categories of products must bear CE marking if you intend to sell them in the European Union (EU) and in member states of the European Economic Area (EEA). Examples of products that require CE marking include:

* electrical products
* construction products
* pressure vessels
* telecommunications equipment
* medical devices
* machinery, equipment and safety components
* personal protective equipment
* gas appliances
* measuring instruments
* lift machinery
* marine equipment.

Note that, where an item of equipment is covered by more than one directive, it must be CE marked under all applicable directives.

Figure 1.3 The CE mark

Environmental health

Local authority regulatory services play a vital role in protecting and supporting the public, the environment and community groups. Local authorities have a statutory duty to enforce a range of environmental regulations that align with national and local policy, codes and guidance.

The impact of decisions taken by Environmental Health Officers (see earlier) can have significant impact on the business concerned and so enforcement actions have to be carefully considered. For example, a low-cost manufacturing process might generate high levels of pollution, necessitating enforcement by the local authority.

Replacing the process with one that is less polluting might have the effect of reducing profitability to a level at which the production is no longer economically viable.

Test your knowledge 1.5

As a result of several serious accidents, a local authority Inspector carries out an inspection of a manufacturing company. What could happen if the inspector discovers a significant failure to safeguard health and safety? Explain your answer.

Test your knowledge 1.6

Explain the role of a Trading Standards Officer.

Test your knowledge 1.7

List **four** examples of manufactured products that require CE marking for sale in the European Union (EU).

Reporting of accidents and injuries

Under the Reporting of Injuries, Diseases and Dangerous Occurrences Regulations 1995 (RIDDOR), employers are required to report a wide range of work-related incidents, injuries and diseases to the Health and Safety Executive (HSE), or to the nearest local authority environmental health department. RIDDOR applies in the following situations:

- work-related accidents which cause death or result in serious injuries
- diagnosed cases of certain industrial diseases
- dangerous occurrences with the potential to cause harm.

The regulations require an employer to record in an accident book the date and time of the incident, details of the person(s) affected, the nature of their injury or condition, their occupation, the place where the event occurred and a brief note on what happened. The report is designed to inform the appropriate enforcing authorities (such as the HSE, local authorities and the Office for Rail Regulation) about deaths, injuries, occupational diseases and dangerous occurrences, so they can identify where and how risks arise, and whether they need to be investigated. This helps the enforcing authorities to target their work and provide advice about how to avoid work-related deaths, injuries, ill health and accidental loss.

For the purposes of RIDDOR, an accident is a separate, identifiable, unintended incident that results in physical injury. Not all accidents need to be reported. A RIDDOR report is required only when the accident is work-related and it results in an injury of a type that's reportable such as a fracture (other than to fingers, thumbs and toes); loss of an arm, hand, finger, thumb, leg, foot or toe; permanent loss or reduction of sight; crush injuries leading to internal organ damage; serious burns (covering more than 10% of the body); problems with the respiratory system or other vital organs.

Injuries to non-workers (e.g. visitors or members of the public) must be reported if a person is injured and is taken from the scene of the accident to hospital for treatment as a consequence. There is no requirement to establish what hospital treatment was actually provided, and no need to report incidents where people are taken to hospital purely as a precaution when no injury is apparent.

Reportable dangerous occurrences

Reportable dangerous occurrences are certain, specified near-miss events with the potential to cause harm. Not all such events require reporting. There are several categories of dangerous occurrences that are relevant to most workplaces. For example, the collapse, overturning or failure of load-bearing parts of lifts and lifting equipment; plant or equipment coming into contact with overhead power lines; explosions or fires causing work to be stopped for more than 24 hours. Additional categories of dangerous occurrences apply to mines, quarries, offshore workplaces and railways.

Recording

A record must be kept of any accident, occupational disease or dangerous occurrence that requires reporting under RIDDOR. These records are important because they ensure that organizations collect sufficient information to allow them to properly manage health and safety risks. This information is a valuable management tool that can be used as an aid to risk assessment, helping to develop solutions to potential risks. In this way, records also help to prevent injuries and ill health, and control costs from accidental loss.

Depending upon the circumstances, different types of report should be made to the HSE. Reports can be made using online forms covering reporting of:

- injuries
- dangerous occurrences
- injuries offshore

- dangerous occurrences offshore
- diseases
- flammable gas incidents
- dangerous gas fittings.

A typical injury report is shown in Figure 1.4.

	Health and Safety Executive

Report of an injury

Note: this is a preview of your form and does NOT represent the submitted details of your notification, which will include the Notification number for reference

About you and your organisation

Notifier name	WORKSHOP MANAGER		
Job title	DAVE CROWE		
Organisation name	SUSSEX CONTROL AND ROBOTICS		
Address	UNIT 4 BROOKES LANE INDUSTRIAL ESTATE		
Phone no	01413 833123	Fax Number	01413 833124
Email Address	DCROWE@SUSSEXCONTROLS.COM		

Where did the incident happen
About the incident

Incident Date	9.3.2016	Incident Time	14.30
In which local authority did the incident occur (Country, Geographical Area and Local Authority)?			
WEST SUSSEX COUNTY COUNCIL			
In which department or where on the premises did the incident happen?			
WORKSHOP STORES			
What type of work was being carried out (generally the main business activity of the site)?			
ENGINEERING MANUFACTURE			

About the kind of accident

Kind of accident	BROKEN LEG
Work process involved	METAL STORE
Main factor involved	COLLAPSE OF SHELVING
What happened	HEAVY LOAD CAUSED SHELVING TO COLLAPSE

About the injured person

Injured persons name	PAUL STEVENS		
Injured persons address	3 GROVE PARK, BILLINGSHURST, RH14 3EL		
Phone no	01413 825571	What was their occupation or job title?	WORKSHOP TECHNICIAN
Gender	MALE	Age	21
Work Status	FULL-TIME EMPLOYED		

About the injured person's injuries

Severity of the injury	REQUIRED HOSPITALISATION		
Injuries	FRACTURE	Part of the body affected	LEG (BELOW KNEE)

Figure 1.4 A typical injury report.

Test your knowledge 1.8

How does RIDDOR define the term 'accident'?

Key point

RIDDOR makes it a legal requirement for organizations to report work-related accidents that cause death or result in serious injuries as well as dangerous occurrences with the potential to cause harm.

Key point

Accident records are important because they ensure that organizations collect sufficient information to allow them to properly manage health and safety risks.

Test your knowledge 1.9

Give **two** examples of near-miss incidents that should be reported under RIDDOR as 'dangerous occurrences'.

Test your knowledge 1.10

Refer to the accident report shown in Figure 1.4 and use it to answer the following questions:

1. On what date and at what time did the accident occur?
2. What was the name of the injured person?
3. What injuries were sustained?
4. What was the job role of the injured person?
5. Who made the accident report and what was their job title?

Activity 1.2

Obtain a copy of '*Reporting accidents and incidents at work A brief guide to the Reporting of Injuries, Diseases and Dangerous Occurrences Regulations 2013*' (available online from the Health and Safety Executive website at www.hse.gov.uk) and use it to answer the following questions:

1. What is RIDDOR?
2. What does RIDDOR apply to?
3. In what circumstances should a report be made and who should make it?
4. Give **four** examples of major injuries that are reportable under RIDDOR.
5. Give **four** examples of dangerous occurrences that are reportable under RIDDOR.

Present your answers in the form of a handout that can be used as part of an A4 'fact sheet' for engineers and managers in an engineering company.

Learning outcome 1.3

State the key duties of the employee in conforming with health and safety requirements

Under Sections 7 and 8 of the Health and Safety at Work Act 1974, it is the duty of the employee to take all reasonable care for their own health and safety as well as that of any other persons who may be affected by their acts and omissions. Section 7 places important duties on the employee, irrespective of the employer's obligations that you will meet in Learning outcome 1.4. It is also worth noting that Section 7 is intended to protect not only fellow employees but also any other person who may be affected by an employee's actions (or lack of action). In the wording used in the Act, Employees are required to:

- co-operate with the employer to enable the duties placed on the employer to be performed
- have regard of any duty or requirement imposed upon their employer or any other person under any of the statutory provisions
- not interfere with or misuse anything provided in the interests of health, safety or welfare in the pursuance of any of the relevant statutory provisions.

In practical terms this means that, as an employee you must take reasonable care of your own health and safety by:

- avoiding wearing jewellery or loose clothing if operating machinery
- if you have long hair, or wear a headscarf, making sure it is tucked out of the way when operating machinery
- taking reasonable care not to put other people – fellow employees and members of the public – at risk by what you do (or don't do) in the course of your work
- co-operating with your employer, making sure you get proper training, particularly in relation to the health and safety aspects of your work
- understanding and adhering to the company's health and safety policies
- not interfering with or misusing anything that has been provided for your health, safety or welfare
- reporting any injuries, strains or illnesses you suffer as a result of doing your job
- telling your employer if something happens that might affect your ability to work, like becoming pregnant or suffering an injury. In an extreme case, and because your employer has a legal responsibility for your health and safety, they may need to suspend you while they find a solution to a health problem, but you will normally be paid if this happens.
- if you drive or operate machinery, telling your employer if you are taking medication that makes you drowsy or impairs your vision. In most cases they will be able to move you to another job on a temporary basis.

Key point

Under the Health and Safety at Work Act 1974, employees have a statutory obligation to ensure the reduction of hazards, ill health and accidents arising from work activities. This means that they should:

- follow appropriate systems of work laid down for their safety
- make proper use of equipment provided for their safety
- co-operate with their employer on health and safety matters
- inform the employer if they identify hazardous handling activities or working conditions
- take care to ensure that their activities don't put others at risk.

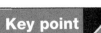

Key point

It is your legal responsibility to take reasonable care for your own health and safety. The law expects you to act in a responsible manner so as not to endanger yourself, other workers, or the general public.

Figure 1.5 As an employee you are responsible for your own safety as well as the safety of those around you.

Test your knowledge 1.11

List **four** practical ways that will help you take care of your own health and safety when working in an engineering workshop environment.

Learning outcome 1.4

State the key duties of an employer in the management of health and safety

Under Section 2 of the Health and Safety at Work Act 1974 it is the duty of the employer to ensure, so far as is reasonably practicable, the health, safety and welfare at work of all employees. The employer also needs to ensure that all plant and systems are maintained in a manner that ensures they are safe and without risk to health. The employer is also responsible for:

- the absence of risks in the handling, storage and transport of articles and substances
- instruction, training and supervision to ensure health and safety at work
- the maintenance of the workplace and its environment to be safe and without risk to health
- providing, where appropriate, a statement of general policy with respect to health and safety, and to provide, where

appropriate, arrangements for safety representatives and safety committees

- conducting their undertakings in such a way so as to ensure that members of the public (i.e. those not in their employment) are not affected or exposed to risks to their health or safety
- giving information about such aspects of the way in which they conduct their undertakings to persons who are not their employees as might affect their health and safety
- the general public (including clients, customers, visitors to engineering facilities and passers-by).

Risk assessment

Under the Management of Health and Safety at Work Regulations 1999 employers have a statutory duty to carry out assessments of the risks that might affect the health and safety of employees and then to act upon the risks they identify. The regulations also require employers to:

- have an effective health and safety policy
- appoint one or more competent persons to oversee health and safety in the workplace (see Section 1.1)
- provide employees with relevant information and training.

Risk assessment will be discussed more fully in Chapter 2.

Shared responsibility

Importantly, it is the duty of *each person* who has control of premises, or has access to or from any plant or substance in such premises, to take all reasonable measures to ensure that they are safe and without risk. This applies equally to employers and employees and both have a hand in ensuring that risks and hazards in the workplace are kept to an absolute minimum.

Key point

The Health and Safety at Work Act 1974 makes both the employer and the employee responsible for safety. Both can be prosecuted for violations of safety regulations and procedures.

Key point

Under the Health and Safety at Work Act 1974, employers have a statutory obligation to ensure the health, safety and welfare at work of all employees.

Test your knowledge 1.12

List **four** practical ways that an employer can help to safeguard the health and safety of employees working in an engineering workshop environment.

Health and safety policies

No matter what size a company is, it should have arrangements, procedures and rules in place to make sure that accidents are prevented during the normal course of its business. Where a company has five or more employees this policy must be written down in the form of a formal statement that sets out its general approach and commitment, together with the arrangements that have been put in place for managing health and safety. It should clearly identify responsibilities within the company for relevant aspects of health and safety. In short, it should say who does what, when and how.

To ensure that the policy is effective and that it takes into account any changes within the organization (including working practices, materials and processes), the policy should be reviewed on a regular basis. Many companies do this at least on an annual basis and, when they do it, they should involve staff at all levels within the organization (not just management).

A health and safety policy is usually made up of several parts, including:

- statements of intent (i.e. what the organization intends to do)
- a list of those responsible for implementing the policy and their roles within the organization
- arrangements and actions (i.e. how specific hazards will be controlled and risks minimized).

Statements of intent indicate the organization's commitment to health and safety and they can be quite brief. For example, 'Prevent accidents and cases of work-related illness' and 'Provide clear instructions, information and training'. The personnel responsible for ensuring that these things happen should be clearly identified. This ensures that people are accountable for what they have to do. Some responsibilities will be fairly obvious. For example, a Workshop Manager will normally be responsible for the maintenance of the plant and equipment within the workshop for which he or she is responsible. Action and arrangements will include a list of specific ways in which health and safety policy is implemented, including:

- risk assessment
- consultation with employees
- safe use of tools, plant and equipment
- safe handling and use of substances
- information, instruction and supervision
- training, induction and updating
- accident, first aid monitoring
- evacuation, emergency and fire procedures.

Case study: Sussex Control and Robotics

Sussex Control and Robotics is a small engineering company that supplies custom-designed control systems and robots used by several large UK manufacturing companies. The company employs a total of 25 staff of which 22 are based at their manufacturing base in West Sussex. The company is managed by its founder Donald Manley who is assisted by Pam Edwards in the day-to-day running of the company. Dave Crowe, the Workshop Manager, is responsible for the plant and equipment which includes advanced CNC machinery and microelectronic assembly.

The company operates from a modern factory unit which incorporates a design office with a small CAD suite. The company employs a receptionist, three customer service support advisers, four sales representatives, two accounts assistants, two design engineers, eight production engineers, three CNC operators, and four field engineers who spend most of their time on the road visiting customers' premises. The office is open Monday to Friday 9.00–5.30 and Saturday morning 9.00–12.00. Office and workshop area cleaning services are provided by a local contractor.

The Manager, Donald Manley, prepared the company's health and safety policy statement based on a template provided by the Health and Safety Executive. He decided to involve his Assistant Manager, Pam Edwards, and Workshop Manager, Dave Crowe, in the production of the policy document (see Figure 1.6) as they both have expertise in health and safety and have thus been given specific areas of responsibility. In line with the policy, the company provides regular training relating to health and safety and this has been designed so that it can also involve the four field engineers who work away from the main company site.

Activity 1.3

Refer to the health and safety policy document shown in Figure 1.6 and use it to answer the following questions:

1. Who has overall responsibility for health and safety within the company?
2. Who is responsible for routinely consulting staff on health and safety matters?
3. Who is responsible for the maintenance of plant and equipment?

Sussex Control and Robotics		Statement of general health and safety policy and arrangements
Donald Manley – Manager (with overall responsibility for health and safety)		
Pam Edwards – Assistant Manager (with day-to-day responsibility for ensuring that this policy is put into practice)		
Statement of general policy	*Responsibility of:*	*Action/Arrangements*
Prevent accidents and cases of work-related ill health by managing the health and safety risks in the workplace	Donald Manley (Manager)	Relevant risk assessments completed and actions arising out of those assessments implemented. (Risk assessments reviewed when working habits or conditions change.)
Provide clear instructions and information, and adequate training, to ensure employees are competent to do their work	Pam Edwards (Assistant Manager)	Staff and subcontractors are given necessary health and safety induction and provided with appropriate training (including working at height, asbestos awareness and electrical safety) and personal protective equipment. We will ensure that suitable arrangements are in place to cover employees engaged in work remote from the main company site.
Engage and consult with employees on day-to-day health and safety conditions	Donald Manley (Manager) Pam Edwards (Assistant Manager) Dave Crowe (Workshop Manager) All staff	Staff are routinely consulted on health and safety matters as they arise but also formally consulted at regular health and safety performance review meetings or sooner if required.
Implement emergency procedures – evacuation in case of fire or other significant incident.	Donald Manley (Manager)	All escape routes are well signed and kept clear at all time. Evacuation plans are tested regularly and updated as necessary.
Maintain safe and health working conditions, provide and maintain plant, equipment and machinery, and ensure safe storage/use of substances	Dave Crowe (Workshop Manager)	Toilets, washing facilities and drinking water provided. Systems are in place for routine inspections and testing of equipment and machinery and for ensuring that action is promptly taken to rectify any defects. Hazardous materials and substances are clearly marked and stored in a secure area.

Health and safety law posters displayed at:	Workshop reception, staff rest room, stores
First-aid boxed are located at:	Main reception, workshop reception
Accident book is located at:	Main reception (accidents and ill health at work reported under the Reporting of Injuries, Diseases and Dangerous Occurrences Regulations)

Signed: **D. Manley**   Date: **18/3/2016**

Figure 1.6 Sussex Control and Robotics health and safety policy statement.

4. What **two** special precautions are observed in relation to hazardous materials and substances?

5. Where is the company's accident book located?

6. Which **four** key health and safety topics are included within the company's induction training?

7. Are subcontractors given any form of health and safety training and, if so, who is responsible for providing it?

8. Under what circumstances are the company's risk assessments reviewed?

Learning outcome 1.5

Recognize the content and application of key health and safety legislation

Employees and their employers need to be fully aware of the need for any organization to meet the relevant health and safety legislation and regulations. So before you get started on developing your engineering skills it is essential to have an understanding of the statutory regulations and safety rules and how they might apply to you. Later you will put this knowledge to good use as you begin to practice your skills and experience some real engineering activities.

In the UK the most important health and safety legislation is:

- The Health and Safety at Work Act 1974
- The Management of Health and Safety at Work Regulations 1999
- The Workplace (Health, Safety and Welfare) Regulations 1992
- The Health and Safety (Display Screen Equipment) Regulations 1992
- The Personal Protective Equipment at Work Regulations 1992
- The Manual Handling Operations Regulations 1992
- The Provision and Use of Work Equipment Regulations 1998
- The Reporting of Injuries, Diseases and Dangerous Occurrences Regulations 1995 (covered on page 13)
- The Working Time Regulations 1998
- The Control of Substances Hazardous to Health Regulations 2002
- The Health and Safety (First-Aid) Regulations 1981
- The Health and Safety (Safety Signs and Signals) Regulations 1996.

There is a lot to take in here so we shall just briefly describe the key aspects of each of these regulations:

The Health and Safety at Work Act 1974

All work activities are covered by the Health and Safety at Work Act 1974 and we have already mentioned this important legislation in relation to the key duties of employers and employees. The Act seeks to promote greater personal involvement coupled with the emphasis on individual responsibility and accountability.

You need to be aware that the Health and Safety at Work Act applies to *people*, not to premises. The Act covers all employees in all employment situations. The precise nature of the work is irrelevant, as is its location. The Act also requires employers to take account of the fact that other persons, not just those that are directly employed, may be affected by work activities. It also places certain obligations on those who manufacture, design, import or supply articles or materials for use at work to ensure that these can be used safely and do not constitute a risk to health.

The Management of Health and Safety at Work Regulations 1999

The duties of employers under these regulations include making assessments of risks that might affect the health and safety of employees and then to act upon the risks they identify. The regulations also require employers to have an effective health and safety policy, to appoint one or more competent persons to oversee health and safety in the workplace, and to provide employees with relevant information and training.

The Workplace (Health, Safety and Welfare) Regulations 1992

These regulations require employers to provide and maintain adequate lighting, heating, ventilation and workplace. Employers must also provide appropriate staff facilities, including toilets, and areas for washing and refreshment.

The Health and Safety (Display Screen Equipment) Regulations 1992

These regulations relate to employees who regularly use display screens as part of the normal daily work. Employers are required

Figure 1.7 Under the Display Equipment Regulations employers are required to carry out risk assessments of CAD workstations.

to carry out a risk assessment of workstations and reduce any risks that they identify. Employers are also required to ensure that users take regular breaks and have regular eyesight tests. There is also a requirement to make furniture (such as workstation desks and chairs) adjustable. Display screen users must also be provided with information on how to recognize and avoid repetitive strain injury (RSI).

The Personal Protective Equipment at Work Regulations 1992

Employers must ensure that suitable personal protective equipment (PPE) is provided free of charge wherever there are risks to health and safety that cannot be adequately controlled in other ways. The PPE must be appropriate and should include items such as protective face masks and goggles, safety helmets, gloves, air filters, ear defenders, overalls and protective footwear, where required. Employers are also required to provide information, training and instruction on the use of this equipment.

Key point

Employers must ensure that suitable personal protective equipment (PPE) is provided free of charge wherever there are risks to health and safety that cannot be adequately controlled in other ways.

Figure 1.8 Under the Personal Protective Equipment at Work Regulations 1992 employers are required to provide appropriate personal protective equipment such as hard hats, gloves and safety glasses.

The Manual Handling Operations Regulations 1992

These regulations require employers to avoid (so far as is reasonably practicable) the need for employees to undertake any manual handling activities that might involve risk of injury. The regulations require employers to make assessments of manual handling risks so that such risks are reduced. Assessments should consider the task, the load and the individual's personal characteristics (physical strength, etc.). Employers should also provide employees with information on the weight of any load that they are expected to lift or move.

The Provision and Use of Work Equipment Regulations 1998

Employers are required to ensure the safety and suitability (i.e. *fitness for purpose*) of work equipment, such as machines and tools. Employers should also ensure that the equipment is properly maintained (regardless of how old it is) and provide appropriate information, instruction and training on its use. There is also a need to ensure that employees are protected from dangerous parts of machinery by the fitting of gates, barriers, guards and shields.

Figure 1.9 Under The Provision and Use of Work Equipment Regulations 1998 employers must ensure the safety and suitability of equipment such as machine tools.

The Working Time Regulations 1998

Fatigue and tiredness not only affects judgement but it is often a contributing factor to workplace accidents and near-miss situations. The Working Time Regulations 1998 implement two European Community directives on the organization of working time and the employment of workers under 18 years of age. The regulations cover the right to annual leave and to have rest breaks, and they limit the length of the working week. Importantly, employers have a contractual obligation not to require an employee to work more than an average 48-hour week (unless the worker has opted out of this voluntarily and in writing).

Key point

Fatigue and tiredness not only affects judgement but it is often a contributing factor to workplace accidents and near-miss situations.

The Control of Substances Hazardous to Health Regulations 2002

Some engineering processes involve materials and substances that can potentially cause harm. The Control of Substances Hazardous to Health 2002 (COSHH) regulations apply to the identification, marking, handling, storage, use and disposal of such substances. Employers must provide appropriate personal protective equipment (PPE) and training in its use. First aid and emergency equipment must be made available and any harmful waste products must be disposed of safely and with consideration for the environment. Information on the storage, use, handling and disposal of hazardous substances is usually provided in the form of safety data sheets. These must be made available for reference in the workplace.

Figure 1.10 Under the Control of Substances Hazardous to Health Regulations 2002 employers must ensure that hazardous materials are labelled and stored securely.

Key point

Information on the storage, use, handling and disposal of hazardous substances is usually provided in the form of safety data sheets.

The Health and Safety (First-Aid) Regulations 1981

The Health and Safety (First-Aid) Regulations 1981 set out the essential aspects of first aid that employers have to address.

The regulations apply to all workplaces, including those with fewer than five employees. As a minimum, a low-risk workplace such as a small office should have a first aid box and a person appointed to take charge of first aid arrangements, such as calling the emergency services when necessary. Employers must also provide information about first aid arrangements to their employees. Workplaces where there are more significant health and safety risks (such as virtually all engineering companies) are more likely to need a trained first-aider. A first aid needs assessment will help employers decide what first aid arrangements are appropriate for their workplace.

The Health and Safety (First-Aid) Regulations 1981 were amended and updated in 2013 so that the HSE no longer needs to approve first aid training or qualifications. The responsibility for this has now shifted onto employers who must satisfy themselves that first aid providers are able to provide appropriate first aid training.

The Health and Safety (Safety Signs and Signals) Regulations 1996

The Health and Safety (Safety Signs and Signals) Regulations 1996 require employers to provide safety signs where other methods cannot deal satisfactorily with certain risks and where the use of a sign can further reduce that risk. They cover traditional safety signs such as 'No entry' signs, and other means of communicating health and safety information such as hand signals, acoustic signals (e.g. warning sirens on machines) and verbal communications such as pre-recorded evacuation messages.

The law harmonizes with an EU directive designed to standardize safety signs throughout member states of the European Union (EU) so that wherever a particular safety sign is seen it provides the same message. The intention is that workers moving from site to site, such as field service engineers, will not be faced with different signs at different workplaces.

In determining when and where to use safety signs, employers must take into account the results of the risk assessment made under the Management of Health and Safety at Work Regulations 1999 (see earlier). This assessment should identify hazards, the risks associated with those hazards, and the control measures to be taken. When those control measures have been put in place there may still be a small but significant risk, in which case employees must be warned of any further measures necessary.

It is important to be aware that the regulations make it clear that safety signs are not a substitute for other means of controlling risks to employees, but they should be used to warn of any remaining significant risk or to instruct employees of the precautions they should take in relation to these risks. It is also worth noting that fire safety signs are covered under separate legal provision.

Figure 1.11 Typical safety signs displayed at the entrance to a construction site.

Test your knowledge 1.13

State the name of the health and safety regulations that relate to:

a) the use of display screens in a computer aided design (CAD) area
b) the maximum number of working hours you can work without taking a break
c) the provision of safety guards around a power guillotine
d) the provision of washroom facilities in a workshop
e) the provision of first aid facilities in a design studio

Activity 1.4

Visit the COSHH section of the UK Government's Health and Safety Executive website (you will find this at www.hse. gov.uk/coshh). View or download a copy of 'Working with substances hazardous to health: A brief guide to COSHH' and use it to produce an A4 'hazard checklist' that can be given to apprentices and new employees in an engineering company. Also include a list of **seven** measures that can be put in place to control the use of hazardous substances.

Review questions

1. Briefly explain the purpose of an inspection carried out by an HSE Inspector.

2. List **four** key tasks performed by a Health and Safety Officer in a small engineering company.

3. Which two key organizations were established by the Health and Safety at Work Act 1974 and what is the relationship between these two organizations?

4. Explain the purpose of an enforcement notice issued by a UK local authority to an engineering company.

5. Explain the purpose of the CE mark.

6. What regulations relate to:
   a) The provision of adequate lighting and ventilation in the workplace?
   b) The use of warning signs in the workplace?
   c) The storage, handling and disposal of hazardous substances?

7. State **four** duties of employers under the Health and Safety at Work Act 1974.

8. Explain what is meant by a 'reportable dangerous occurrence'. Why is it necessary to report such an incident?

Chapter checklist

Learning outcome	Page number
1.1 Recognize the roles of key people involved in workplace health and safety	4
1.2 Recognize the roles of organizations involved in health and safety	8
1.3 State the key duties of the employee in conforming with health and safety requirements	16
1.4 State the key duties of an employer in the management of health and safety	18
1.5 Recognize the content and application of key health and safety legislation	23

CHAPTER **2**

Health and safety in the engineering environment

Learning outcomes

When you have completed this chapter you should understand how health and safety procedures are applied in an engineering environment, including being able to:

2.1 Recognize the procedures in performing a risk assessment activity.

2.2 Implement how to safely perform manual handling tasks.

2.3 Recognize the procedures for working in dangerous circumstances.

2.4 Recognize how to comply with organizational safety requirements.

2.5 Implement fire and emergency evacuation procedures.

Chapter summary

In the previous chapter we introduced specific roles and responsibilities in relation to health and safety in the workplace. We also introduced the legislation and regulations that govern working practices in engineering. This chapter takes this further by explaining how health and safety procedures are applied in an engineering environment. We look in particular at the way that hazards are identified and the procedures that must be adopted when working in dangerous situations such as underground, in confined spaces, in bad weather, or at height.

Learning outcome 2.1

Recognize the procedures in performing a risk assessment activity

As you have already learned in Chapter 1, employers are obliged to control the risks present in a workplace. To do this they need to think about what might cause harm to people and decide whether they are taking reasonable steps to prevent that harm. This is known as *risk assessment* and is something that an employer is required to do by law. The Health and Safety Executive (HSE) recommends five stages in carrying out a risk assessment, namely:

* Identify the hazards that might cause harm.
* Decide who might be harmed, and how they might be harmed.
* Evaluate the risks and decide on precautions.
* Make a record of your findings.
* Periodically review your assessment and update if necessary.

These five stages are often referred to as the *five steps to risk assessment*. Let's look at them in a little more detail:

Step 1: identify hazards that might cause harm

Employers have a duty to assess the health and safety risks faced by their workers. For example, your employer must systematically check for any physical, mental, chemical, electrical and biological hazards that might be present in the workplace.

* *Physical hazards* can involve lifting, awkward postures, slips and trips, noise, dust, machinery, computer equipment, etc.
* *Mental hazards* can result from excessive workload, working with difficult clients, working under pressure, bullying, etc.
* *Chemical hazards* can arise from the presence of dangerous

substances such as asbestos, fuels and lubricants, cleaning fluids, aerosols, etc.

* *Electrical hazards* can involve shocks and burns from exposed wiring and high voltage cables.
* *Biological hazards* are rare in an engineering context but can arise from untreated water supplies, electromagnetic radiation from X-ray equipment, radar sets and transmitting equipment, and intense ultraviolet (UV) radiation from arc welding equipment.

An important point to note is that, while many hazards are obvious and can be seen and quickly identified, others cannot. A good example is radiation from X-ray equipment which is invisible to the eye. In addition to all of this, fire can result in burns and respiratory problems caused by smoke and fumes. We will look at fire procedures separately at the end of this chapter.

Figure 2.1 Many different hazards can be present in an engineering environment.

Step 2: decide who may be harmed, and how they might be harmed

Anyone present in an engineering environment should be considered. People who might be at risk may include others as well as those directly involved with carrying out engineering tasks, such as visitors, contractors, trainees, delivery drivers and maintenance staff.

Employers must review work routines and practices in all of the locations and situations where staff are employed. For example, in offices, store rooms, reception areas, plant rooms and gangways as well as workshops and production areas. Employers also have a special duty of care towards the health and safety of young workers, disabled employees, night workers, shift workers, and pregnant or breastfeeding women.

Step 3: assess the risks and take action

Employers must consider how likely it is that each hazard could cause harm. This will determine whether or not an employer should take steps to reduce the level of risk. Even after all precautions have been taken, some risk will usually remain and employers must decide for each remaining hazard the level of risk and whether it is high, medium or low. This is not an easy process and requires judgement by the person(s) carrying out the risk assessment.

Step 4: make a record of the findings

Employers with five or more staff are required to record in writing the main findings of the risk assessment. This record should include details of any hazards noted in the risk assessment, and action taken to reduce or eliminate risk. The written record provides proof that the assessment was carried out, and should then be used as the basis for a later review of working practices. The written record should make reference to other health and safety regulations where relevant. For example, if fire safety signs are not prominently displayed in an area where naked flames are present.

Step 5: review the risk assessment

A risk assessment must be kept under review in order to ensure that agreed safe working practices continue to be applied (e.g. that safety instructions are understood and are respected) and also to take account of any new working practices, new machinery or more demanding work targets.

Key point

Risk assessment is about identifying and implementing measures that can be introduced to control *risk* (or the chance that somebody could be harmed by a particular hazard). When carrying out a risk assessment, remember that a hazard is anything that could cause injury such as working at height, working with toxic materials, etc.

Case study: Aircom Masts and Towers

Aircom is a specialist manufacturer of masts and towers used in the general aviation sector. The company supplies sectional and pneumatic masts capable of supporting headloads of up to 300 kg at a height of over 30 m. The masts are typically used to support antennae and lighting systems for small airfields and for temporary use at exhibitions and air shows. The company's masts and towers are suitable for ground mounting or can be deployed by means of a four-wheel trailer which can be towed by a conventional road vehicle.

Aircom employs a total of 15 staff located at their manufacturing base with access from the A40 at Oxford Airport where examples of the company's products are in use. The company recently opened a new production workshop adjacent to their main site. This will be used for assembling trailers supplied in kit form by a manufacturer in South Wales. Once assembled, each trailer needs to be adapted so that a mast or tower can be fitted to it.

Equipment in the new workshop comprises a small car lift (used to raise each trailer to a convenient working height), a bench grinder, arc welding equipment, a pillar drill and various 240 V operated portable tools.

The Production Manager, Steve Ellis, and Safety Officer, Phil Green, recently carried out a risk assessment of the new trailer workshop and a brief extract from it is shown in Figure 2.2. During the risk assessment various observations were made and actions noted. Following on from the assessment it was felt that some changes should be made to the company's induction training for new staff and that there is a need for a regular review of some of the workshop's PPE.

Key point

Don't forget that a risk assessment is a working document. You should be able to read it and return to it, noting actions that have been taken and identifying any new hazards and risks that might have appeared as a result of changes to equipment, materials, personnel or working practices.

AIRCOM MASTS AND TOWERS		Risk assessment for: *Welding Bay in Trailer Assembly Workshop (WR1)*				
Unit 2, Spitfire Way, Oxford Airport, OX9 3SD		Activities carried out on this site: *Arc welding*				
		Date of this assessment: *17/03/16*				
What are the hazards?	**Who might be harmed and how?**	**To control this risk:**			**Action:**	
(Give full details)	(Give full details)	What is already being done?	What additional action is needed?		By who?	By when?
Arc welding; intense UV radiation, ozone and nitrogen dioxide (generated by UV light), sparks and flying particles of slag, fire (due to malfunction of electrical equipment and flying particles of slag), production,	Welding Technicians; short-term photoconjunctivitis ('arc eye'), cataract (long-term), retinal injury (long-term); short-term erythema, blistering and burns to exposed skin (short-term); cancer (long-term) Production staff in adjacent areas and occasional workshop visitors; short-term photoconjunctivitis ('arc eye')	A written Safe Operating Procedure is provided. This is clearly displayed at the entrance The Welding Bay is closed off from the main workshop and is fitted with fume extraction. Hazard and safe working practice signs are clearly displayed. PPE (including gloves, aprons, full-face visors) is available and in good condition. Visors are compliant with BS EN 169 and BS EN 175	The Safe Operating Procedure should be included in the induction training given to all production workers A notice should be displayed to indicate that the Welding Bay is a restricted area and that access is limited solely to Welding Technicians Visors should be periodically checked for wear and damage. Any damaged eyewear should be withdrawn from service immediately		Phil Green Steve Ellis Phil Green	Before next induction training 01/04/16 Monthly

Figure 2.2 An extract from the Aircom Masts and Towers risk assessment.

Test your knowledge 2.1

List each of the 'five steps to risk assessment'.

Test your knowledge 2.2

Explain briefly how the level of a particular risk is evaluated when a risk assessment is being carried out.

Activity 2.1

Carefully read Aircom's risk assessment for the welding bay in their new trailer assembly workshop (Figure 2.2) then use it to answer the following questions:

1. When was the risk assessment carried out?
2. What **three** additional actions were identified and who is to be responsible for them?
3. What is 'arc eye' and why is it a problem?
4. What is UV and why is it a hazard? What measures have been implemented to control it?
5. What is ozone and why is it a hazard? What measures have been implemented to control it?
6. Apart from UV radiation, what other hazards have been identified?
7. What **two** potential causes of fire have been identified?
8. Explain why production staff in nearby areas and occasional visitors are not likely to be harmed by burns.

Activity 2.2

Carry out a detailed risk assessment of the electrical/electronic workshop of your workplace, training centre or college. Your risk assessment must be based on the activities that are actually carried out in the workshop so it is a good idea to start off by finding out who uses the workshop and what processes are in use.

Based on your risk assessment, present your findings in the form of a ten-minute presentation using appropriate visual aids and with supporting notes classifying risks as 'high', 'medium' and 'low'. The following questions should help you to get started:

1. Where is the electrical circuit breaker? (It should be in a prominent and immediately accessible position.)
2. Has a residual current circuit breaker (RCCB) been fitted and, if so, has the RCCB been tested lately?
3. Are all of the electrical outlets in a safe condition?
4. Does each item of equipment have a mains lead and plug that is in a safe condition?
5. Is each item of test equipment correctly fused?
6. Have portable items of electrical equipment been tested in accordance with Portable Appliance Testing (PAT) regulations? If so, when was the last PAT test carried out?

7. Is the soldering equipment safe? Are soldering irons fitted with heatproof leads and is there any provision for fume extraction?
8. Are safety glasses available? What condition are they in and where are they stored?
9. Is the lighting adequate?
10. Are the safety exits properly marked and unobstructed?
11. Are appropriate fire extinguishers available and when were they last tested?
12. Is appropriate safety information on display?

Your presentation should include recommendations on the safety of materials and equipment handling, use of personal protective equipment and the potential hazards that you have identified in the area. You should also suggest ways in which a work activity (such as PCB etching, bench drilling or soldering) could be made less hazardous (for example, by improving the lighting or the use of fume extraction equipment).

Learning outcome 2.2

Implement how to safely perform manual handling tasks

The Manual Handling Operations Regulations 1992 define manual handling as '... *any transporting or supporting of a load (including the lifting, putting down, pushing, pulling, carrying or moving thereof) by hand or bodily force*'. In an engineering context the load can be almost anything that needs to be moved, such as metal stock, parts and materials. By virtue of their shape, many loads can be awkward to move even though they might not be considered very heavy. For example, a large sheet of aluminium can be relatively light but at the same time it can be difficult to grip and move in the restricted space of a workshop. It also has sharp edges and corners that constitute an additional hazard when being moved.

The Manual Handling Operations Regulations 1992 suggest a three-stage approach:

1. Avoid hazardous manual handling operations as far as is reasonably practicable.
2. Assess any hazardous manual handling operations that cannot be avoided.
3. Reduce the risk of injury so far as is reasonably practicable.

Assessing manual handling risks

It is sometimes difficult to assess the risk associated with performing a particular manual handling operation. This is not only because people are physically different (height, weight, etc.) but also because the nature of the load and the circumstances in which it is moved might vary. Moving a large mechanical engineering component such as a motor or pump might be relatively straightforward in a well-lit air-conditioned workshop but would be altogether more problematic if the move had to be performed on the platform of a North Sea oil rig during a winter storm.

An effective way of assessing manual handling activities involves looking at four key factors; Task, Individual, Load and Environment (the word 'TILE' will help you to remember this). By considering each of the TILE factors separately you will be able to make a realistic assessment of the risk and its severity:

1. *Task* – what does the activity involve and does it involves working in a confined space, bending, stooping, twisting, etc.? Is the load stationary and can it be gripped easily? Are there any other factors associated with the task that might increase the severity of risk?

2. *Individual* – does the person (or persons) performing the task possess normal physical strength or are they restricted in any way (perhaps by height or reach)? Are they 'able bodied' or do they suffer from any form of disability or medical problem that might impact on their ability to perform the task? Have they received any specific training that might be needed in order to perform the task? Do they have previous experience of performing the task and how recent is this experience?

3. *Load* – is the load heavy, unbalanced, unwieldy, difficult to grasp, sharp, hot, cold or difficult to get hold of? Is the load clearly marked with information or restrictions on its handling (for example, 'do not stack' or 'this way up')? Is the load fragile or breakable? What would happen if the load is dropped? Would there be any spillage of fuel, lubricant or other material that might then constitute a hazard?

4. *Environment* – is the area accessible and does it have sufficient space in which to move the load? Is the floor level and free from obstructions? Is the floor damp, oily or slippery? Is the lighting adequate and is it uniform (with no shadows or 'glare spots')? Is wind gust or airborne pollutants (such as smoke or chemical fumes) a problem? If PPE is required does this restrict movement or reduce breathing and/or visibility in any way?

Duties of employers

In relation to manual handling, employers need to comply with the risk assessment requirements set out in the Manual Handling Operations Regulations 1992 (in addition to that required by the Management of Health and Safety at Work Regulations 1999). To assist employers with this task, specific guidance is given within the manual handling regulations. Employers should also consult and involve their employees, particularly those with first-hand experience of carrying out manual handling tasks. In many cases, employees will be able to offer practical solutions for controlling hazards and minimizing risk.

Duties of employees

Employees have duties to take reasonable care of their own health and safety and that of others who may be affected by their actions. This means that, in addition to general health and safety duties, in relation to manual handling they must:

- follow appropriate systems of work laid down for their safety
- make proper use of equipment provided for their safety
- co-operate with their employer on health and safety matters
- inform the employer if they identify hazardous handling activities
- take care to ensure that their activities do not put others at risk
- make use of appropriate handling aids
- put into practice training they have received in relation to manual tasks.

Test your knowledge 2.3

List the four key 'TILE' factors that need to be considered when assessing a risk associated with a manual handling task.

Lifting and moving a load

First of all, it is important to know why you are moving a heavy load and whether it really is necessary to move it. For example, if you are applying a particular process to a heavy workpiece would it be easier to move the process rather than the workpiece? Secondly, rather than use manual handling could you be using mechanical aids such as a conveyor, pallet truck, trolley, a powered hoist, a winch or a forklift truck? Also, will you need assistance in order to help manage the load or perhaps to provide guidance and direction?

A second opinion and a second pair of hands can be invaluable in many situations. What PPE will you need? Gloves, hard hats

Figure 2.3 Correct technique for lifting a heavy or bulky object.

Figure 2.4 Incorrect technique for lifting a heavy or bulky object.

> ### Key point
> Correct manual handling is essential. When lifting heavy and bulky objects the straight back, bent legs technique should be used.

and safety boots or shoes might be needed in a variety of manual handling operations. Having the right PPE can be instrumental in controlling risks when moving a heavy, bulky or awkward load.

Despite this, lifting a heavy and/or bulky object is a task that most engineers will perform from time to time. In order to avoid accidents and injuries, engineers need to adopt safe working practices and, in many cases, this will involve training in manual handling techniques.

Correct lifting technique

Using correct lifting technique can be instrumental in avoiding back and other muscle injury. The technique is as follows:

1. Plan the lift in advance. Where is the load now and where is it to be placed? Can lifting aids be used? What PPE is required? How heavy is the load and are there any restrictions on the way that it can be positioned or handled? When moving a heavy or unwieldy load some distance you might need to rest the load on an appropriate support. This will provide you with an opportunity to rest and, if necessary, change your grip.
2. Adopt a stable position. This usually involves positioning yourself with feet slightly apart facing the object to be lifted and at a comfortable distance from it (about half an arm's length is usually ideal).
3. Bend your legs (not your back) and grasp the object at each side with a firm grip (use handles where provided).
4. Keeping your back straight unbend your legs slowly and evenly, raising the object to the most comfortable height. Avoid twisting or leaning. Ensure that you keep your head up and that you have good visibility while moving the load.
5. Lower the load into the position that you have previously prepared (ensuring adequate clearance). If necessary, the final position can be adjusted after the load has been put down and made safe.

Figures 2.3 and 2.4 illustrate the correct and incorrect lifting technique. When placing a heavy load on a table, bench or other work surface it is essential to ensure that the object can be adequately supported and that it will not slip or fall. With some objects (for example, those that are heavy, slippery, or have sharp edges) it is essential to use protective clothing such as overalls, gloves, and safety boots or shoes. In addition, hard hats are obligatory wherever overhead work is being carried out and also when you are working underground or on construction sites. The maximum recommended manual handling load weights at specified heights and distance from the body are shown in Figure 2.5 for both men and women. Note how the maximum load is significantly reduced as the distance from the body increases.

Test your knowledge 2.4

What part of the body is most likely to be injured using the incorrect lifting technique shown in Figure 2.4 and why does the technique shown in Figure 2.3 minimize this risk?

Activity 2.3

Visit the Health and Safety Executive website at http://www. hse.gov.uk. Locate and download a copy of *'Manual handling at work: A brief guide'*. Use it to answer the following questions:

1. What is an 'MSDS'?
2. List **five** problems to look for when making an assessment of the working environment.
3. List **three** topics that should be included in a manual handling training programme.
4. Is there a difference between what people can lift and what people can safely lift? Explain why this is important.
5. What is the maximum recommended weight of a load that can be lifted at waist height and at arm's length by a man? Does this increase or decrease as the load is placed a) above the chest or b) nearer the body?

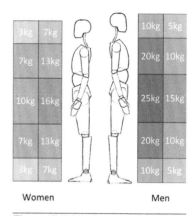

Women   Men

Figure 2.5 Maximum load weight at specified heights and distances.

Key point

Maximum load weight is significantly reduced as distance from the body increases.

Learning outcome 2.3

Recognize the procedures for working in dangerous circumstances

Engineering activity can take place in a huge variety of different environments. Whilst hazards can exist in all environments they can be more severe in situations involving confined spaces, trenches, at height, with chemicals/toxic substances, dust-enriched atmospheres or damp/wet atmospheres.

Confined spaces

A confined space is a place which is partly or completely enclosed, and where there is a risk of death or serious injury from hazardous substances or dangerous conditions (such as a lack of oxygen). Confined spaces can include storage tanks, silos, reaction vessels, enclosed drains and sewers, open-topped chambers, containers, battery rooms, ductwork and poorly ventilated rooms.

Confined spaces can be deadly and a number of people are killed or seriously injured each year in the UK working in a confined space. This happens in a wide range of different industries, including oil and gas, construction and manufacturing.

There is an added danger associated with confined spaces. If a worker is overcome with fumes (or as a result of a lack of oxygen), a person sent in to attempt a rescue is also liable to be overcome for the same reason unless adequate precautions are taken (such as the use of breathing apparatus). You can find more information in the Confined Spaces Regulations 1997 which specify the following key duties:

- as far as possible, avoiding entry to confined spaces (e.g. by doing the work from outside)
- following a safe system of work if entry to a confined space is unavoidable
- putting in place adequate emergency arrangements before the work starts.

Working at height

Working at height can be equally challenging but the hazards are quite different. Working at height invariably requires appropriate PPE, including hard hats and safety harnesses. Ladders and platforms need to be stable before use, and hand rails and guard rails need to be in place to assist with movement and prevent falls.

Chemicals and toxic substances

All hazardous materials should be clearly marked and stored safely and securely. Material Safety Data Sheets (MSDS) should be available for each hazardous material. These are available from manufacturers and suppliers and can often be downloaded from the internet. The MSDS should contain a brief description of the material or substance, together with comprehensive information about the product including details relating to handling, use, storage, transportation and disposal. Details of the PPE required to work safely with the product should also be included. Anyone working with hazardous materials should have access to the MSDS for the materials concerned.

Atmospheric pollution

Face masks (either full or partial) will be required in an environment with a polluted atmosphere. In an extreme case breathing apparatus may be required. In all cases specialist advice will be necessary in order to know what specific pollutants are present and how they can best be controlled.

Figure 2.6 Dangerous and toxic substances need to be clearly marked and stored safely and securely.

Test your knowledge 2.5

What is an MSDS and in what circumstances would you need one?

Activity 2.4

Visit a website such as ScienceLab.com (http://www.sciencelab.com) and search for the MSDS for ferric chloride test solution. Use the MSDS to answer the following questions:

1. What is the chemical composition of the substance and what is its physical appearance?
2. In what way(s) is the substance hazardous?
3. What first aid measures should be applied in the event of eye contact?
4. What first aid measures should be applied in the event of skin contact?
5. Is the substance flammable or explosive?
6. What precautions should be observed when storing the material?

Learning outcome 2.4

Recognize how to comply with organizational safety requirements

In order to deal with specific hazards, safe working procedures (SWP), safe operating practice and standard operating procedures (SOP) are used in many industries as well as in many engineering companies. Despite the different names, these documents are basically similar and usually take the form of a step-by-step description of how a process is performed, taking into account hazards that might be present and the control measures needed to reduce risk.

An SWP or SOP usually lists the hazards involved in performing a task, what risk is associated with each hazard and an indication of its severity. PPE is invariably identified along with the steps necessary to complete the activity without incident. An SWP or SOP is usually divided into several parts using the same broad headings found in a risk assessment. Various control measures may be identified within an SWP or SOP. Sometimes these are intended to prevent work starting unless it is safe for it to do so (a *permit to work*) or to ensure that a process can't start by physically preventing it (a *lock-off*). We will now look at these in a little more detail.

Isolation

In some cases, it may be necessary to isolate part of a process so that it can be worked on without potential interference and hazards caused by adjacent processes. This is often the case when maintenance, repair or equipment replacement becomes necessary.

In order to perform most maintenance tasks, guards, barriers and enclosures will need to be removed. If this is the case, additional measures will be needed to prevent danger from the mechanical, electrical and other hazards that may be exposed. There should be clear company rules on what isolation procedures are required, and in what circumstances they should be used.

Another important control measure is that there should be isolation from the power source (usually, but not exclusively, electrical energy), the isolator should be locked in position (for example by a padlock), and a sign should be used to indicate that maintenance work is in progress. Isolation requires use of devices that are specifically designed for this purpose; not devices such as key-lockable emergency stops or other types of switches that may be fitted directly to a machine. Any stored energy (hydraulic or pneumatic power, electric charge, etc.) must be safely dissipated before the work starts.

Lock-off

If more than one person is carrying out work at the same time then, having isolated the supply each person should be able lock off the power using their own padlock. In this situation, removing just one of the locks will not allow the supply to return; they must all have been removed before the supply can be reinstated. Multi-padlock hasps can be used in such circumstances. Similar isolation procedures can be applied to locking off valves for services such as steam, liquids and gases. Before entering or working on the equipment, it is essential that the effectiveness of the isolation is verified by a suitably competent person.

Permits to work

A permit to work is a formal, written, safe system of work to control potentially hazardous activities. The permit details the work to be done and the precautions to be taken (for instance, they might involve closing down part of a process or disconnecting supplies, or they might detail rescue arrangements for certain types of work). Permits should be issued, checked and signed off as being completed by a competent employee (often a site manager) who is not directly involved in carrying out the work.

Permits to work will tend to be appropriate in the following situations:

- where contractor's work interrupts normal production activities
- work on plant where fumes, toxic or radioactive materials, steam or gases may be present
- hot work which could cause fire or explosion
- entry into fuel or waste tanks, containers, battery rooms and other confined spaces.

Case study: North Western Aggregates

North Western Aggregates operates a number of quarries in Cumbria and Lancashire where they extract gravel and stone for use in construction and road building. The gravel is passed through a crushing plant where it is reduced in size to the required grade. The company employs 85 staff but finds it cost-effective to use local contractors to maintain most of its heavy plant equipment. Since the processes associated with gravel extraction are hazardous (particularly in the crushing plant) there is need to ensure that contractors follow appropriate safety procedures and don't put themselves or others at risk when they are working on any of the company's sites.

To maintain high safety standards, North Western Aggregates makes extensive use of permits to work. These must be obtained and approved before any on-site work can be carried out by contractors. The permits to work must be generated by a company supervisor and they must be approved by a site manager or quarry manager.

When a contractor arrives on-site, he or she is required to report to the site office and show the signed permit to work. This is then recorded in the site visitor's book. Work may then start, subject to the agreed safety procedures and precautions stipulated in the

<table>
<tr><td colspan="2">

NORTH WESTERN AGGREGATES PERMIT TO WORK

</td></tr>
<tr><td>Site or quarry: HEYWOOD MOOR</td><td>Valid from: 22-3-16</td></tr>
<tr><td>Plant or zone: CRUSHING PLANT B</td><td>Valid to: 23-3-16</td></tr>
<tr><td colspan="2">Contractor: HALL AND PLATT</td></tr>
<tr><td colspan="2">Supervisor: HUGH JONES</td></tr>
<tr><td colspan="2">Named operative(s): TONY HALL
GREGG PHILLIPS</td></tr>
<tr><td colspan="2">Work for which Permit to Work is required:
INSPECTION AND MAINTENANCE OF AIR FILTERS</td></tr>
<tr><td colspan="2">Required safety precautions (including PPE):
CLOSE DOWN AND LOCK-OFF ALL SUPPLIES
HARD HATS, FACE MASKS, EYE PROTECTION AND GLOVES
NB: MUST MARK AREA WITH 'CONTRACTOR AT WORK' SIGN</td></tr>
<tr><td colspan="2">Safety planning/risk assessment completed?   Yes [X]   Pending [ ]   No [ ]</td></tr>
<tr><td>Approval of Permit to Work:

NB: Only Site Manager or Quarry Manager to approve</td><td>Signed   Scott Parker
Date   14 March 2016</td></tr>
<tr><td>Acceptance of Permit to Work:</td><td>Signed   A. J. Hall
Date   18/3/16</td></tr>
<tr><td>Completion of Work:
ENSURE WASTE MATERIAL IS REMOVED!</td><td>Signed
Date</td></tr>
<tr><td>Renewal of Permit to Work:
Approved until</td><td>Signed
Date</td></tr>
<tr><td colspan="2">**IMPORTANT**: This Permit to Work **MUST** be shown to the Site Office on arrival and retained for inspection when required. Completed Permits to Work must be returned to the Site Office **BEFORE** leaving the site. If you need further information please call 01452 779 977 or 01339 452444.</td></tr>
</table>

Figure 2.7 Permit to work issued to Hall and Platt (contractors) by North Western Aggregates.

Key point

To reduce hazards and minimize risk when inspecting, maintaining or repairing plant and equipment, it is usually necessary to isolate the supply of electricity, gas or other fuel, etc. Where more than one person is working at the same time it might also be necessary to lock off the supply.

Key point

Permits to work are frequently required where contractors are working on a site and where normal production activities are taking place. Permits to work are usually only effective for the time needed to complete the work and they will normally specify the PPE required as well as any special safety precautions.

permit to work. When the work is complete the permit to work should be returned to the site office where it is filed in case it is needed for later reference.

Test your knowledge 2.6

What is a safe working procedure (SWP) and what would you expect to find in it?

Test your knowledge 2.7

What is a permit to work and in what circumstances might you need one?

Activity 2.5

Carefully read the permit to work issued to Hall and Platt (Figure 2.7) then use it to answer the following questions:

1. On what specific date was the permit to work operative?
2. Which two employees of the contractor are named in the permit to work?
3. What items of PPE are specified in the permit to work?
4. What additional precaution has been identified as necessary before work can start?
5. Who has approved the permit to work and when was it approved?
6. On completion of the work what additional task has been specified by the company supervisor?
7. What should happen if the work cannot be completed by the date specified?
8. Suggest **two** reasons why the permit to work is filed in the site office after the work has been completed.

Learning outcome 2.5

Implement fire and emergency evacuation procedures

Fire is the rapid oxidation (burning) of flammable materials. For a fire to start, the following are required (see Figure 2.8):

Figure 2.8 The fire triangle.

- a supply of flammable materials
- a supply of air (oxygen)
- a heat source.

Once the fire has started, the removal of one or more of the above will result in the fire going out.

Fire prevention

Fire prevention is largely a matter of common sense and 'good housekeeping'. The workplace should be kept clean and tidy. Rubbish should not be allowed to accumulate in passages and disused storerooms. Oily rags and waste materials should be put in metal bins fitted with airtight lids. Plant, machinery and heating equipment should be regularly inspected, as should fire alarm and smoke detector systems. Electrical wiring, alterations and repairs must only be carried out by qualified electricians and must comply with the current Institution of Engineering and Technology (IET) Regulations. Smoking must be banned wherever flammable substances are used or stored. The advice of the fire prevention officer of the local brigade should be sought before flammable substances, bottled gases, cylinders of compressed gases, solvents and other flammable substances are brought on-site.

Test your knowledge 2.8

List the **three** factors that need to be present for a fire to start.

Activity 2.6

Visit the Health and Safety Executive (HSE) website at www.hse. gov.uk and obtain information on safety in the gas welding and cutting process. Identify at least **five** main hazards associated with the process and explain what a backfire and a flashback is. Present your answer in the form of a brief written advice notice to be displayed in the welding area of an engineering workshop.

Safety and warning signs

Appropriate warning and prohibition signs should be prominently displayed wherever necessary. The five main types of sign, shown in Figure 2.9, are as follows:

- Prohibition signs (things that you *must not* do, for example, 'No Smoking')

Figure 2.9 Different types of sign.

- Mandatory signs (signs that indicate things that you *must* do, for example, 'Eye Protection must be used')
- Warning signs (signs that warn you about something that is dangerous, e.g. 'Danger High Voltage')
- Safe condition signs (signs that give you information about the safest way to go, for example, 'Fire Exit')
- Fire signs (signs that indicate the location of fire equipment, for example, 'Fire Point').

Note that different colours are used to make it easy to distinguish the types of sign. For example, safe condition signs use white text on a green background, mandatory signs use white text on a blue background, and so on. It is essential that you familiarize yourself with the different types of sign and know what they mean!

Test your knowledge 2.9

Classify each of the signs shown in Figure 2.10 as either a prohibition sign, a mandatory sign, a warning sign, a safe condition sign, or a fire sign.

Fire extinguishers

Several different forms of fire extinguisher are provided in order to cope with different types of fire:

Figure 2.10 Safety and warning signs.

1. Class A extinguishers are for ordinary combustible materials such as wood, paper, cardboard and most plastic materials. This type of extinguisher is based on water and may involve the use of a wall-mounted reel and hose.
2. Class B extinguishers are used for fires involving flammable or combustible liquids such as fuels, solvents, oil and grease.
3. Class C extinguishers are used where fires involve electrical equipment, wiring, switchgear, circuit breakers and outlets (water-based extinguishers should **never** be used on this type of fire).
4. Class D extinguishers are suitable for use with chemical fires involving combustible metals such as magnesium, titanium, potassium and sodium.

It is essential to remember that, whilst water can be an effective extinguishing agent for Class A fires (paper, wood, etc.), water and air-pressurized water (APW) extinguishers **must not** be used on Class B or Class D fires because their use can actually cause the flames to spread and make the fire bigger! Dry powder extinguishers come in a variety of types and are suitable for a combination of class A, B and C fires. These are filled with foam or powder and pressurized with nitrogen. Carbon dioxide (CO_2) extinguishers are designed for use with Class B and C fires. CO_2 extinguishers contain carbon dioxide, a non-flammable gas, and are highly pressurized. Fire blankets (typically of woven fibreglass) can also be used to extinguish small fires by smothering and cutting off the supply of oxygen to the fire.

It is worth remembering that, unlike dry chemical powder extinguishers, CO_2 extinguishers don't leave a harmful residue. This can be important when dealing with fires on expensive electrical and electronic equipment which may suffer permanent damage when dry powder extinguishers are used. Finally, it is important to have appropriate training in the operation of fire extinguishers and other fire protection equipment. If your company, training centre or college can provide you with this training, you should take full advantage of it!

Emergency instructions and evacuation procedures

Engineering companies will have in place a set of instructions and evacuation procedures that should be followed in the event of fire, explosion, chemical or fuel spillage. These instructions should be prominently displayed and all employees need to be familiar with them. Where visitors or contractors are on-site they too need to be aware of the evacuation instructions. Instructions and procedures should include:

1. The action that employees should take if they discover a fire (or other serious emergency situation). This will usually involve operating the nearest alarm call-point and attacking the fire, if possible, with appliances immediately available, but without taking personal risks.
2. Checking that the fire alarm has been raised and that all nearby personnel are alerted to the situation.
3. Information on how to leave the building by the nearest protected route (this must be clearly marked with prominently placed signs) or via a designated emergency exit. Note that lifts should never be used.
4. Arrangements for the safe evacuation of people identified as being especially at risk, such as contractors, those with disabilities, members of the public and visitors. Special equipment, such as safety slides and fire refuges, may be provided for those suffering from disabilities that could prevent them from leaving a building safely using stairs.
5. Reporting to a designated assembly point where a designated *fire warden* or *fire marshal* will conduct a roll-call of those present, ensuring that none are unaccounted for.
6. Identifying the type and location of fire equipment. Normally, fire extinguishers will be conspicuously located in circulation areas (such as the base of a stairwell) and near fire exit doors.
7. The duties of employees with specific responsibilities in the event of fire (such as fire wardens and fire marshals) will also be

Figure 2.11 A typical fire point with CO_2 and dry powder extinguishers.

Figure 2.12 An aircraft hangar fire exit clearly marked with safe condition signs.

stated. These might include the need for fire marshals to ensure that their allocated areas are clear of personnel and that visitors are accounted for with reference to the visitors' book (note that host employees should normally be responsible for their visitors, ensuring that they are escorted safely from the site). Designated employees will be responsible for calling the fire brigade and also for liaison on-site, notifying fire officers of any specific hazards that might be present such as fuel tanks, chemical stores, etc.

8. Details of training, including the frequency and conduct of fire drills, the testing of fire alarms, and regular training for fire wardens and fire marshals, as appropriate. A record of training and attendees should be kept for inspection when required.

Activity 2.7

Figure 2.13 shows the fire instructions that appear in Aircom's new trailer workshop. Read the fire instructions and answer the following questions:

1. Where is the nearest exit?

2. Where is the assembly point?

Key point

It is essential to get to know the emergency procedures for evacuation, accident, fire, etc. in your training centre or workplace. You need to know where the fire exits, alarms and fire points are and how to get assistance from a first-aider or fire warden.

Figure 2.13 Aircom's fire instructions displayed in their new trailer workshop.

3. What should employees do when they arrive at the assembly point?

4. What should employees not do?

5. Under what circumstances should employees tackle a fire?

Test your knowledge 2.10

State which type of fire extinguisher you would use in each of the following cases:

a) Paper burning in an office waste bin.

b) A pan of fat burning in the kitchen of the works canteen.

c) A fire in a mains voltage electrical machine.

Test your knowledge 2.11

A fire breaks out near to a store for paints, solvents and bottled gases. What action should be taken and in what order?

Test your knowledge 2.12

List **six** key points that should be included in an engineering company's fire instructions.

Review questions

1. Explain why it is necessary to carry out a periodic review of a risk assessment.

2. Give **three** examples of physical hazards present in a typical engineering environment.

3. Give **three** examples of biological hazards present in a typical engineering environment.

4. List **four** factors that need to be considered when planning to lift a heavy object manually.

5. When using manual handling, explain why maximum load weight must be reduced as distance from the body increases.

6. Briefly explain the responsibilities of a fire warden or fire marshal.

7. The Manual Handling Operation Regulations 1992 suggest a three-stage approach to manual handling. What are the three stages?

8. List **four** statutory duties of employers in relation to manual handling.

9. Where would you expect to find detailed information on a hazardous chemical substance?

10. What is a permit to work and why would you need one?

11. Give an example of a) a safe condition sign and b) a warning sign. In what way do these signs differ?

12. List **three** different types of fire extinguisher and give an example of where each type would be used.

Chapter checklist

Learning outcome	Page number
2.1 Recognize the procedures in performing a risk assessment activity	32
2.2 Implement how to safely perform manual handling tasks	37
2.3 Recognize the procedures for working in dangerous circumstances	41
2.4 Recognize how to comply with organizational safety requirements	44
2.5 Implement fire and emergency evacuation procedures	47

CHAPTER **3**

Safe moving and storing of materials

Learning outcomes

When you have completed this chapter you should understand how health and safety procedures are applied in an engineering environment, including being able to:

3.1 Describe the range and applications of equipment available to assist in moving loads correctly and safely.

3.2 Recognize how to safely move loads.

3.3 Recognize how to correctly store gases, oil, acids, adhesives and engineering materials.

Chapter summary

Most engineering activities and processes involve moving and storing materials. This chapter introduces you to the equipment and techniques used to safely lift, transport and store a wide range of different materials. When moving and storing materials, safe working practices are essential in order to eliminate the risk of damage and injury. This is particularly important when working with hazardous materials such as fuels, gases, solvents and chemicals.

Learning outcome 3.1

Describe the range and applications of equipment available to assist in moving loads correctly and safely

In the previous chapter we described the techniques for manual handling and lifting. While this might be appropriate for everyday lifting operations in a practical engineering context you will undoubtedly encounter some loads that will exceed the lifting capability of the average human being. In such circumstances you will need to make use of one or more lifting aids, such as bars, rollers, pallet trucks, forklifts and cranes. As an engineer you need to be familiar with this equipment even though you may not be using any of it on a regular basis.

Figure 3.1 Engineers using various lifting aids to remove an aircraft engine.

Pinch bars and rollers

Pinch bars (*crowbars* or *pry bars*) consist of a simple lever bar with a shaped end (the *toe*) that can be inserted under the base of a load. The bar can then be levered so that the edge of the load is lifted, providing access to the underside of the load so that rollers or skates can be inserted to assist movement.

Hydraulic toe jacks

Hydraulic toe jacks can be used in very confined spaces in order to further reduce the lifting effort. Typical uses include lifting and positioning of heavy machinery, steel structures and general engineering applications. Hydraulic toe jacks are ideal for use with load moving *skates* which support the load whilst it is being moved or transported.

Pallets

Pallets are frequently used for transporting loads. They aid handling by providing a firm and stable platform on which a load can be supported. They also make it possible for lifting aids such as pallet trucks and forklifts to be used without the need to raise the load off the ground. Most pallets are flat and open above the base but some are available with frames and cages.

Stillages

Stillages (or *cage pallets*) are a special type of pallet that incorporates a frame and cage or some other means of containing and supporting a load. Stillages are commonly used to transport

Figure 3.2 A stillage or cage pallet.

goods safely and without damage. They also simplify the process of loading and unloading large numbers of products or transferring them from a vehicle to a warehouse. Another common use for stillages is as a bin for accumulating material, such as rejected parts or scrap metal. Stillages are usually fitted with feet to make them self-supporting and also to aid in forklifting and stacking.

Pallet trucks

Pallet trucks are one of the most common ways of transporting loads through short distances. They are highly manoeuvrable and offer an excellent lifting solution for both light and heavy loads. Most pallet trucks are designed for use with standard European, UK and American pallets as well as pallets from the Far East. Pallet trucks are suitable for lifting loads of up to several tonnes. Pallet trucks are fitted with hydraulic pumps that permit raising and lowering of the forks. Rollers, either single or tandem, fitted to the forks help spread the load on the floor surface. More sophisticated pallet trucks are available that will raise a load to heights in excess of 1.5 metres. Others are motorized to reduce the effort required when moving.

> **Key point**
>
> Pallet trucks provide a cost-effective way of moving and transporting loads through short distances.

> **Key point**
>
> Stillages provide a means of containing a load within a cage so that it can be lifted and moved using a pallet truck.

Figure 3.3 A conventional pallet truck can be safely used to move a stillage.

Test your knowledge 3.1

Describe a stillage and explain how it is used.

Scissor lifts

Scissor lifts employ hydraulic scissor action to raise a working platform to a considerable height. They are available in various sizes and heights ranging from around 5 to 15 metres. Scissor lifts are often used for indoor construction, maintenance and installation applications. A scissor lift is a safer and more stable solution for working at height than using a ladder and offers greater flexibility than scaffolding.

Figure 3.4 Scissor lifts are used to elevate platforms to the required working height.

Forklift trucks

Forklift trucks have two forks (similar to a pallet truck) a motor drive with rear wheel steering and seating for an operator. The rear wheel drive helps to improve manoeuvrability, particularly in tight cornering situations. However, since the centre of gravity of a forklift varies with the load mass and its position, forklift trucks are inherently unstable. Because of this it is important to avoid making turns at speed or with raised loads.

The load limit of a forklift decreases with fork height and a manufacturer's loading plate with reference data is usually located in a prominent position. A wide variety of different types of forklift is available including those suitable for applications such as loading and unloading goods, line feeding, stacking, order picking, and for general movement of loads up to about 8 tonnes. It should go without saying that a forklift truck should never be used for lifting people!

>
>
> **Key point**
>
> Forklifts are rated for a specified maximum load and forward centre of gravity. This information can usually be found on a plate provided by the manufacturer.

Figure 3.5 Forklift trucks are ideal for loading and unloading goods.

Test your knowledge 3.2

Explain why a forklift should not be driven round tight corners when the load is raised.

Cranes

Various types of crane are used in engineering and manufacturing. They include wall, overhead, jib and tower types capable of lifting and moving bulky and heavy loads. Various factors need to be considered when selecting the type of crane best suited for use in a particular application, including the weight and dimension of the load, the height of lift and distances/areas of movement required, the number of lifts required and the time available, and the ground/floor conditions. Being able to operate a crane requires specialist training.

Fixed *overhead cranes* (or *bridge cranes*) are frequently used in large-scale manufacturing environments where they can handle loads of up to around 100 tonnes. An overhead crane consists of parallel rails with a bridge that travels along the rails, spanning a gap over which the load can be moved. The lifting part of the crane is a hoist fitted to the travelling part. The rails of the bridge are supported by the structural walls of the building in which the crane is used. Alternatively, if the bridge is self-supported, the crane is called a *gantry* or *goliath* crane. Unlike mobile jib cranes, overhead cranes are permanent and available for use whenever required.

This makes them more cost-effective for use in manufacturing and maintenance applications.

Wall cranes and *pillar cranes* are similar to overhead cranes but they are anchored to a wall or pillar at one end only. Instead of a fixed set of rails, they have a *jib* rail that can slew over a wide angle in the horizontal plane. This makes them suitable for accurately positioning loads of up to 10 tonnes over a relatively large area.

Figure 3.6 An overhead crane in a large engineering workshop.

Figure 3.7 Controls for the overhead crane in Figure 3.6.

Ancillary equipment

A variety of different equipment is available to assist with moving and lifting loads including, block and tackle, ratchet hoists, slings, shortening clutches, lifting/plate clamps, eye bolts and shackles of various types. All of these lifting aids are designed to ensure that the load is attached safely and securely. All lifting equipment should be tested by qualified inspection personnel and clearly marked with its safe working load (SWL).

Block and tackle

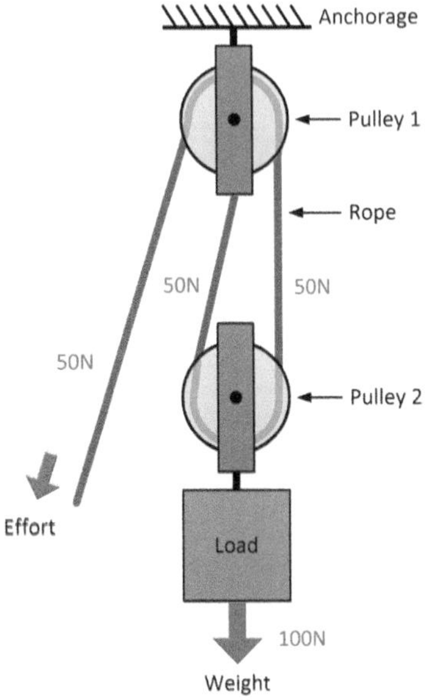

A block and tackle hoist provides a simple but effective way of reducing the effort required to raise a load and consists of a system of two or more pulleys with a rope or cable threaded between them. The pulleys are enclosed in blocks that are paired with one of them free to move whilst the other is anchored to a support. The system provides *mechanical advantage* so that the applied force is a fraction of the weight force to be lifted.

The mechanical advantage, M, of a pulley system is given by the relationship:

$$M = \frac{\text{Load (or \emph{weight})}}{\text{Effort (or \emph{pull})}} = n$$

where n is the number of rope sections linking the two pulleys.

In the simple block and tackle hoist shown in Figure 3.8 one pulley is anchored and the other is attached to the load. There are two rope sections linking the two pulleys and thus the mechanical advantage is 2. If the weight force exerted by the load is 100 N the effort (pull) required to lift the load (ignoring friction) would be half this, or 50 N. Note that the tension force in each of the rope sections is 50 N (half the weight force). Notice also that, when raising the load, the effort needs to move through twice the distance that the load moves.

Figure 3.8 A simple block and tackle arrangement providing a mechanical advantage of 2.

> **Key point**
>
> All lifting equipment should be tested by qualified inspection personnel and clearly marked with its safe working load (SWL).

> **Key point**
>
> The mechanical advantage of a pulley system reduces the force required to raise a load.

Test your knowledge 3.3

What is 'mechanical advantage' and why is it important?

Ratchet hoists

Ratchet hoists (also known as *pull lifts*) are versatile lifting aids that use a chain operated by a hand lever and ratchet mechanism. They can usually be operated in any orientation and they are suitable for almost any pulling, lifting or hoisting application with a typical load capacity of up to several tonnes. Ratchet hoists have an integrated brake system which holds the load. Direction may be controlled by a simple three-position lever marked up, down and neutral (see Figure 3.10). When in the neutral position the operator can easily adjust the chain length before tensioning the load.

Figure 3.9 A ratchet hoist in use.

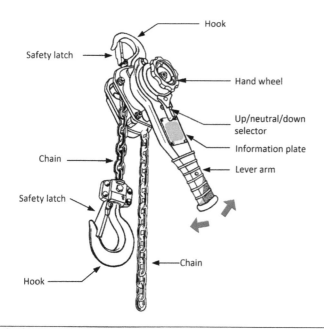

Figure 3.10 Adjustment points on a typical ratchet hoist.

Key point

Pull-lifts comprise a chain and ratchet mechanism operated by a hand lever. These tools are attached using hooks fitted with safety latches and they are capable of lifting loads of up to several tonnes.

Test your knowledge 3.4

State **two** advantages of a ratchet hoist when compared with a simple arrangement of pulleys.

Test your knowledge 3.5

Describe the adjustments provided on a typical lever-operated ratchet hoist.

Slings

Slings are used to attach a load to a lifting device such as a crane or hoist. Depending on the size and physical nature of the load, various types of sling can be used. Slings may typically be attached to the load at one, two, three or four points. Slings are frequently made from metal chains but they can also be made from rope and polyester webbing.

Chain slings tend to be stronger, more reliable and robust than other types. They can maintain their safe working loads over a wide range of temperatures but they suffer from the disadvantage that if one link fails the entire chain becomes unsafe. Defects can be difficult to spot from a quick visual inspection and chain slings can suddenly break without warning. In contrast, damage and defects can be easier to spot with a rope sling and failure tends to be gradual rather than sudden.

Wire rope consists of individual wires laid into a number of strands, which are then wrapped around a central core. Wire rope slings are less expensive than chain slings and they can be made in different lengths where several attachment points are necessary. By virtue of their construction, it can be difficult to bend a wire rope around a tight radius bend and so *shackles* are normally employed to attach them to a load. To form a loop, the end of the wire rope is normally fitted with a thimble loop and several (usually three) cable clamps are used to retain the rope.

Polyester webbing slings are often used for loads that can be easily damaged. Unlike chains and wire rope, they are light and easily handled. However, they are susceptible to damage and must be protected from sharp edges and from sliding along the load when used at an angle. Regardless of which type of sling is used, they must be visually checked before each use and quarantine for further inspection if found to be defective in any way.

Figure 3.11 A sling arrangement being used to lift a coil of sheet steel.

Test your knowledge 3.6

Describe **three** different types of sling used for attaching a load to a hoist.

Shortening clutches

Shortening clutches enable a multi-legged chain sling to have different length legs to adjust to the balance or required lifting angle of the load. Clutches must be correctly fitted to avoid failure or shortening of the chain life.

Hooks and end fittings

The end fittings on chain slings (and occasionally rope slings) will generally be either sling hooks fitted with safety latches or C-shaped hooks. Both are designed to minimize the risk of the sling slipping out of the hook. In addition, the C-hook is designed not to catch on obstructions and is more robust as it does not have an easily damaged safety latch. Special-purpose fittings based on clamps and plates are used with some types of load, including pipes and drums.

Eye bolts

In order to make lifting easier and simplify attachment, eye bolts can be fitted to heavy equipment such as motors, generators, alternators or transformers. Simple *dynamo eye bolts* are often considered to be a permanent part of the equipment to which they are attached and they are only suitable for axial loading (in other words along the axis of the screw). This would be the case with a straight vertical lift using a single hook and sling. Although these types of eye bolt have a small collar it is usually insufficient to withstand significant loading applied at an angle to the thread axis.

Collared eye bolts are similar but they have a much larger collar and a relatively small eye opening. This design helps to avoid large stress concentrations which can lead to fatigue failure. They are also able to withstand angled loading which is inevitable when more than one attachment point and several slings are used. Note that although this type of eye bolt is able to withstand side loading, the opening of the eye should always be aligned in the direction of the pull.

Figure 3.12 A sling hook fitted with a safety latch.

Figure 3.13 An eye bolt connected to a heavy load.

Test your knowledge 3.7

Explain the difference between a dynamo eye bolt and a collared eye bolt.

Shackles

Shackles or *hook rings* are divided into two main categories: simple *D-type shackles* or *bow-type* (anchor) *shackles*. Both are available with pins that can be threaded or secured with nuts. Figure 3.14 shows how the basic types of shackle are used and the *crown points* where the load is applied.

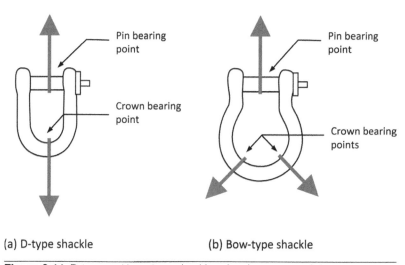

(a) D-type shackle   (b) Bow-type shackle

Figure 3.14 D-type and bow-type shackles showing crown point.

Figure 3.15 A wire rope connected to a D-shackle.

Figure 3.16 End loops of steel rope slings showing thimbles used to define the end shape.

Regulations

Two important regulations impact on the safe lifting and movement of materials. They are the Provision and Use of Work Equipment Regulations 1998 and the Lifting Operations and Lifting Equipment Regulations 1998. We will briefly explain the relevant features of each of them.

The Provision and Use of Work Equipment Regulations 1998 (PUWER)

Any equipment that is used by an employee at work is covered by The Provision and Use of Work Equipment Regulations 1998 (PUWER). This includes equipment such as ladders, drilling machines, power presses, circular saws, photocopiers, lifting equipment (including lifts), dumper trucks and motor vehicles. Note also that, if employees provide their own tools and equipment and if they use them at work, they must also comply with PUWER.

Under PUWER, employers need to ensure that the work equipment is:

* suitable for use, and also suitable for the purpose and conditions in which it is to be used
* maintained in a safe condition for use so that people's health and safety is not at risk
* inspected to ensure that it is (and continues to be) safe for use.

Any inspection should be carried out by a competent person (this can be an employee if they have the necessary skills, knowledge and experience to perform the task). Records of inspection must be retained.

Employers need to ensure that risks created by using the equipment they provide are eliminated wherever possible or controlled as far as reasonably practicable. Typical measures that an employer might take include providing effective guards and protection devices (such as emergency stop buttons), as well as appropriate PPE. In addition, employers should ensure that hazards are clearly marked and warning signs are prominently placed. Safe working procedures need to be in place and adequate information, instruction and training needs to be given to employees. In many cases a combination of these measures will be necessary.

The Lifting Operations and Lifting Equipment Regulations 1998 (LOLER)

The Lifting Operations and Lifting Equipment Regulations 1998 (LOLER) place duties on people and companies who own, operate or have control over lifting equipment. This includes all businesses and organizations whose employees use lifting equipment, whether

> **Key point**
>
> The Provision and Use of Work Equipment Regulations 1998 (PUWER) requires that work equipment and plant should be maintained so that it remains safe and also that the maintenance operation is carried out safely. As well as machinery, tools and plant, this includes equipment used for lifting, moving and storing materials.

owned by them or not. In most cases, lifting equipment is also work equipment so the Provision and Use of Work Equipment Regulations 1998 (PUWER) will also apply (including inspection and maintenance). All lifting operations involving lifting equipment must be properly planned by a competent person, appropriately supervised and carried out in a safe manner.

As with PUWER, LOLER requires that all equipment used for lifting is fit for purpose, appropriate for the task, suitably marked and, in many cases, subject to statutory periodic thorough examination. Records must be kept of all examinations and any defects found must be reported to both the person responsible for the equipment and the relevant enforcing authority.

Key point

The Lifting Operations and Lifting Equipment Regulations 1998 (LOLER) requires that all equipment used for lifting is fit for purpose, appropriate for the task, and has been periodically checked.

Test your knowledge 3.8

Briefly explain the key features of PUWER and LOLER that relate to lifting and moving loads.

Learning outcome 3.2

Recognize how to safely move loads

Safe and correct working practice when moving a load usually involves:

- knowing the safe working load (SWL) of each item of lifting equipment and ensuring that it is not exceeded
- making and following an effective lifting plan
- fitting lifting slings correctly and ensuring that they have a maximum spread of 120°
- ensuring an adequate clearway when the load is moving
- not moving loads over others' heads
- not using lifting equipment to transport people
- using the correct hand signals for crane operators
- protecting chains, ropes and slings from sharp corners
- setting hooks and sling/chain lengths correctly.

Safe working load

The safe working load (SWL) of lifting equipment is the maximum load that the equipment can safely lift. If more than the SWL is applied there is a risk of failure which can be catastrophic, resulting in serious damage to both the load and lifting equipment as well as injury to personnel working nearby. Because of this all lifting equipment, including accessories, must be clearly marked to indicate their SWL. This information should also relate to the different configurations in which the equipment can be used. For example, where the hook of an engine hoist can be moved to different positions, the SWL should be shown for each position.

Sling configuration

The choice of sling configuration for a particular lift depends on a number of factors including:

- the size of the load (larger loads will require longer slings)
- the weight of the load (heavier loads will require appropriately rated slings)
- the shape and orientation of the load
- the nature of the load and whether it can be easily damaged
- whether the load is a single item or a bundle of items.

Two slings are frequently used, in which case the length of the slings and the distance between the two attachment points on the load will determine the internal angle between the slings. This angle is important since the tension force in the slings will increase as their internal angle increases, see Figure 3.17. When two slings are used the internal angle should not normally be allowed to exceed 90° with 120° usually stated as the absolute maximum. Figure 3.18 illustrates this important point. Notice how, with an internal angle of 120°, each of the supporting slings will be subject to a tension force equal to the weight of the load.

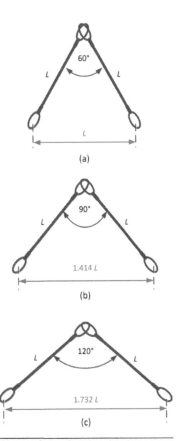

Figure 3.17 Relationship between sling length, L, internal angle, and the distance between attachment points.

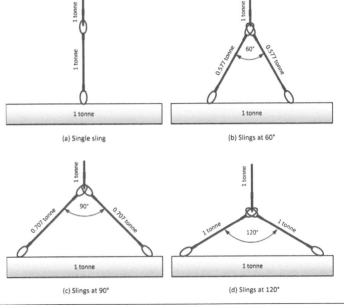

Figure 3.18 Relationship between the internal angle and the sling tension for a 1,000 kg (1 tonne) load.

Test your knowledge 3.9

Estimate the internal angle between the supporting slings shown earlier in Figure 3.11. Explain why this angle should not normally be allowed to exceed 90°.

Activity 3.1

Figure 3.19 shows a 90 kVA diesel generator that needs to be moved a distance of 10 m using a small mobile crane. The generator weighs 2,650 kg. Make a complete list of the lifting equipment required and, with the aid of a sketch, explain how the equipment should be configured. Specify a minimum SWL for each item.

Key point

The configuration of a sling depends on a number of factors including the weight, size and shape of the load.

Key point

The tension force in a sling harness will increase significantly as the internal angle increases. To avoid excessive tension force the internal angle should not normally exceed 90° and in no case should it be allowed to exceed 120°.

Figure 3.19 90 kVA diesel generator.

Crane operator signals

When moving a large load it will often be necessary for directions to be given to the crane operator by someone on the ground with better visibility. This is often done using hand-held radios but when a radio fails or is unavailable a standard set of hand signals should be used (see Figure 3.20).

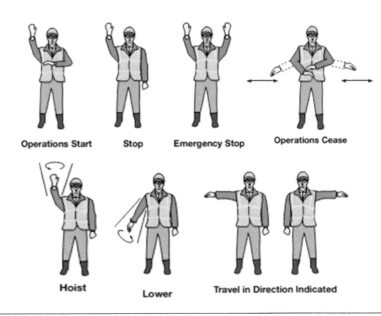

Operations Start   Stop   Emergency Stop   Operations Cease

Hoist   Lower   Travel in Direction Indicated

Figure 3.20 Crane operator signals.

Learning outcome 3.3

Recognize how to correctly store gases, oil, acids, adhesives and engineering materials

Careful consideration needs to be given to the means of storing hazardous materials and substances used in engineering such as gases, lubricants, fuels, acids, solvents and adhesives. Storage needs to be safe, secure and effective and must comply with the Control of Substances Hazardous to Health (COSHH) regulations. Recall from Chapter 1 that the COSHH regulations apply to the identification, marking, handling, storage, use and disposal of hazardous substances. Also that employers must provide appropriate personal protective equipment (PPE) and training in its use. First aid and emergency equipment must be made available and any harmful waste products must be disposed of safely and with consideration for the environment. Information on the storage, use, handling and disposal of hazardous substances should be available for reference in the workplace.

Control measures

When working with hazardous materials (and in order to comply with the COSHH regulations) it is necessary for employers to have in place effective control measures. These should be designed so that they reduce risk by examining each material and process in which it is used as well as the equipment used to control the risk. It also involves the adoption of safe working practices and how management can be sure that the risks really are being controlled. Table 3.1 provides you with four typical examples of control measures used in engineering companies. Notice how the storage of materials and substances features in each of these control measures.

Table 3.1 Some typical control measures used to minimize the risks when working with hazardous materials.

Material/process	Control equipment	Way of working	Management
Solvents applied using cloths and aerosol sprays	Waste materials need to be placed in a marked bin fitted with a lid. Solvents need to be clearly marked and stored securely	Avoid skin contact, use gloves, reduce build-up of vapour using local ventilation	Check controls are used. Aerosol sprays need to be clearly marked and securely stored. Monitor regular disposal of waste material
Beryllium oxide used in high power semiconductors	Defective or replaced devices must be placed in clearly labelled and sealed bags before returning to the manufacturer	Avoid skin contact and inhalation. Wear protective clothing, gloves and full face protection. Where local ventilation is unavailable respiratory protection must be worn	Check controls are used. Minimize use wherever possible. Ensure training is provided for all staff concerned
Cutting fluid mist produced by a lathe	Put an enclosure around the lathe and extract the mist via an air filter. Cutting fluid needs to be stored in a secure area. Used air filters need to be placed in sealed bags before storage	Allow time for mist to clear before opening the enclosure. Use skin care products where contact has occurred	Check and maintain fluid quality. Check that the extraction system is working and that filters are replaced regularly. Ensure training is provided for staff concerned
Lubricating oil used in engine crankcases	Ensure that the oil is clearly marked and stored in a secure compound. Ensure that equipment for filling and draining is effective and working properly	Avoid skin contact. Avoid inhalation of vapour or mist. Wash hands and exposed areas	Check and monitor control measures. Ensure that used oil containers are stored and disposed of safely

Storage areas

Access to hazardous material needs to be strictly controlled and a record needs to be kept of all incoming and outgoing materials. The responsibility for keeping such records will normally be attributed to a store keeper or workshop technician and will be included in his or her job specification.

Strorage areas need to have sufficient space and they need to be well organized, well lit and well ventilated. The store area should have an impervious floor resistant to fluids and chemicals and it should be easy to keep clean. The area should be maintained at a consistently cool temperature and it should be appropriately ventilated.

Incompatible materials should be stored well away from one another. Accidental contact between incompatible chemicals can result in a fire, an explosion, the formation of highly toxic and/or flammable substances, or other potentially harmful reactions. For example, oxidizers mixed with flammable solvents can cause a fire, acids mixed with metal dust can produce flammable hydrogen gas, and so on.

Chemical and fuel storage areas should be arranged so that leaks or spills are controlled. Chemicals and fuels should never be allowed to flow onto floors, into waste water or storm drains. Spills and leaks need to be cleaned up immediately.

Materials should be stored on racks or shelves and they should be clearly marked. Aisles, access ways, hallways, doorways, exits and entryways should be kept clear, and emergency exits should be clearly marked. Shelving should be level, stable and secured to a wall or other stable surface.

Key point

Chemical and fuel storage areas should be arranged so that leaks or spills can be controlled.

Figure 3.21 Fluids stored in racks.

Test your knowledge 3.10

Give two examples of incompatible materials and explain why they should not be stored in close proximity.

Flammable materials

Flammable materials (including lubricants, oil, fuels, adhesives, paints and varnishes) should always be kept well away from ignition sources such as open flames, hot surfaces, direct sunlight, sparks, etc. Flammable materials should also be separated from other hazardous materials. Flammable liquids should only be stored in approved safety containers or cabinets. In instances where static electricity may accumulate and there is a risk of flammable vapours becoming ignited, liquid containers and tanks should be bonded together and grounded by means of a low-resistance electrical connection. Appropriate fire extinguishing equipment should be prominently placed and training for its use provided for all those who work in the area. Appropriate safety signs should be prominently displayed.

Key point

Flammable materials (including lubricants, oil, fuels, adhesives and varnishes) should always be kept away from ignition sources.

Compressed gases

As a result of the compression of a gas, compressed gases possess a large amount of stored energy. Damaging a gas cylinder by dropping or knocking it can cause the energy to be rapidly released. Additional hazards can arise from the flammability or reactivity of

Figure 3.22 Stillages containing gas cylinders.

Key point

Gas cylinders can be dangerous. They should only be stored in racks or frames and kept well away from sources of heat. Cylinders should be handled carefully to avoid knocking or dropping.

the gas. Where several different types of gas are being stored they should be separated and segregated. In addition, the quantity of gas cylinders stored in any particular area should be limited to only those that will be used within a reasonable period of time. Cylinders should always be stored upright and in suitable racks or frames. Cylinders should be kept away from heat and open flames and the valve protection cap on the cylinder should be kept in place whenever the cylinder is not in use. They should never be stored in confined spaces where no ventilation is present. Safety signs should be prominently displayed.

Acids and corrosive materials

Strong acids can destroy human tissue and corrode metals. Acids are incompatible with one another and may react with many other substances. Use tight-fitting goggles, gloves, and closed-toe shoes while handling acids and other corrosive materials. Acids should only be stored in approved non-reactive containers on lower shelves, at least below eye level. Containers should be placed in plastic tubs or trays that will provide secondary containment. Acids and other corrosive materials should not be stored on metal shelves. Although ventilation helps, chemicals will still corrode the shelves. Appropriate safety signs should be prominently displayed.

Test your knowledge 3.11

Explain why secondary containment might be necessary when storing acids and other corrosive materials.

Activity 3.2

A small engineering company needs to establish a store for hazardous materials where lubricants, solvents and battery acid will be stored. Write a checklist of at least **ten** different things that need to be taken into account when planning the design and layout of the storage area.

Activity 3.3

Use the internet and other resources to determine the meaning of each of the safety signs shown in Figure 3.23.

(a)

(b)

(c)

(d)

(e)

Figure 3.23 Safety signs.

Review questions

1. Explain why the safe working load (SWL) of lifting equipment is important.

2. Identify the lifting aid shown in Figure 3.24. Describe a typical application for this device.

Figure 3.24 Lifting aid.

3. Briefly describe the operation of a ratchet hoist.

4. State **two** advantages of a wire rope sling compared with a chain sling

Figure 3.25 Lifting aid.

5. Identify the lifting aid shown in Figure 3.25. Describe a typical application for this device.

Figure 3.26 Lifting aid.

6. Identify the lifting aid shown in Figure 3.26. Describe a typical application for this device.

Figure 3.27 Ancillary item.

7. Identify the ancillary item shown in Figure 3.27. Describe a typical application for this device.

Figure 3.28 Ancillary item.

8. Identify the ancillary item shown in Figure 3.28. Describe a key
    safety feature of this device.

9. List **four** factors that need to be considered when preparing a
    sling for lifting a heavy load.

10. What essential control measures should be taken when storing
    solvents and adhesive?

Chapter checklist

Learning outcome	Page number
3.1 Describe the range and applications of equipment available to assist in moving loads correctly and safely	56
3.2 Recognize how to safely move loads	69
3.3 Recognize how to correctly store gases, oil, acids, adhesives and engineering materials	72

CHAPTER 4

Engineering and the environment

Learning outcomes

When you have completed this chapter you should understand how environmental management is applied in an engineering environment, including being able to:

4.1 Analyse the relevant legislation and EU directives with regard to environmental management.

4.2 Explain what is contained in the environmental management systems BS EN ISO 14001 (EMS).

4.3 Explain what climate change levy (CCL) is and its implications, and what is exempt.

4.4 Describe what sources of energy are available other than fossil fuels.

4.5 Evaluate the criteria with regard to emissions.

4.6 Recognize the requirements for safe disposal of waste.

Chapter summary

Engineering activities can have a significant impact on the environment. However, in the past half century, new laws have been introduced to minimize this impact, consequently making the world a safer, cleaner place. As a result, it has become increasingly important for engineers to understand how, and why, environmental legislation impacts on normal day-to-day engineering activities. We begin this chapter by looking at the key legislation and explaining what it aims to do before moving on to look at ways in which responsible engineering companies deal with waste and unwanted emissions. The Activities introduced in this chapter will bring into sharp focus crucial issues such as the effects of climate change, alternative 'green' energy sources, and how to deal with the accidental release of a polluting substance.

Learning outcome 4.1

Analyse the relevant legislation and EU directives with regard to environmental management

A wide variety of environmental legislation impacts directly on UK engineering activities. You need to understand how and why this has become important and how it affects the day-to-day engineering tasks and activities that you will be involved with. In this first section we will briefly introduce the legislation that you need to be aware of.

The Environmental Protection Act 1990

The Environmental Protection Act 1990 (EPA) is a UK Act of Parliament that aims to control waste and pollution from industrial processes. The EPA deals with issues relating to waste and how it is managed. It applies to anyone who produces, imports, transports, stores, treats or disposes of *controlled waste* and it requires those involved to hold the necessary *authorization* issued by local authorities and the Environment Agency. The EPA also places a duty on local authorities to collect waste. Companies have a *duty of care* to ensure that any waste they produce is handled safely and within the law.

The Pollution Prevention and Control Act 1999

Under the Pollution Prevention and Control Act 1999 (PPCA), operators of specified installations are required to obtain a *permit* issued by the appropriate enforcing authority. Detailed conditions are attached to permits. These relate to the operation of the process, setting limits to minimize the emission of *pollutants* and also requiring regular monitoring of *emissions*.

Under the PPCA the impact of the *installation* is considered as a whole; this includes emissions to the air, water, land and also noise output, waste minimization and energy usage. In permitting an installation, conditions can be applied by the regulator (often the local authority) to the activities carried on to minimize pollution in accordance with the current best practice in the industry (sometimes referred to as the *best available technique*).

An important element in the granting of a permit is the emphasis on the prevention of pollution (which was not a prominent part of the EPA). The operator of an installation (i.e. plant or processing facility) must now be able to demonstrate that the environmental impact of any proposed change on-site has been assessed and controlled. However, as technology changes, the current best practice will also change and so the conditions contained within permits cannot be considered fixed forever. Subsequent review of permits may require operators to adopt different (and more effective) controls as and when they become available. A permit will usually specify:

- the activities taking place within the site or installation
- the pollution controls required to meet the current best available technique
- the monitoring required by the operator to demonstrate compliance.

Key point

In order to control and reduce emissions of waste and pollutants from certain industrial processes, the Environmental Protection Act 1990 (EPA) requires those involved to hold the necessary *authorization* issued by local authorities and the Environment Agency. The Pollution Prevention and Control Act 1999 (PPCA) takes this one stage further by requiring regular monitoring of emissions, setting limits that minimize the emission of pollutants.

Figure 4.1 By law, the operator of an installation must be able to demonstrate that the environmental impact has been assessed and controlled.

Key point

The Pollution Prevention and Control Act 1999 aims at controlling the environmental impact of an industrial or commercial installation.

4

The Clean Air Act

The first UK Clean Air Act was introduced in 1952 as a result of the Great London Smog. At this time the air in London (and many other cities) became very severely polluted by smoke and fog lasting for many days. The legislation sought to control atmospheric pollution by introducing *smokeless zones* in which only smokeless fuels can be burnt. The introduction of cleaner coals and the increased usage of electricity and gas further helped to reduce the level of pollutants in the atmosphere. In addition, there was a move to relocate coal-burning power stations to more rural areas. The result was a significant reduction in air pollution and, in particular, the level of *sulphur dioxide* in the atmosphere.

Further legislation was introduced with the Clean Air Acts 1968 and 1993. The 1968 Act required the construction of taller chimneys for industries burning coal, liquid or gas fuels. Tall chimneys are more effective at dispersing sulphur dioxide which, in 1968, could not be effectively removed from exhaust gases by any practical means. The 1993 Act also introduced a wide range of new regulations including the content and composition of motor fuels and vehicle emissions.

Figure 4.2 Atmospheric pollution over London resulting mainly from exhaust emissions.

The Radioactive Substances Act 1993

The Radioactive Substances Act 1993 deals with the control of radioactive material and disposal of radioactive waste in the UK. These regulations were replaced and amended by the Environmental Permitting Regulations 2010 (EPR). The scope of the EPR now includes water discharge and groundwater activities, as well as radioactive substances and a number of directives concerning batteries and mining waste.

The Environmental Permitting Regulations 2010

Under the Environmental Permitting Regulations 2010 (EPR), standard rules apply to the issue of *permits* to carry out specific activities. This simplifies the process of obtaining a permit but the operator is obliged to comply with a fixed set of standard rules designed to ensure that the installation does not cause pollution, nuisance or damage to the environment.

The Controlled Waste Regulations 2012

Because of its hazardous nature, toxicity or the possibility of harm to human health or the environment, some waste needs to be treated with special care. This type of waste is strictly controlled and it is referred to as *controlled waste*. A prime concern with controlled waste is the effects of biodegradation or biochemical degradation and the by-products that might be produced. The Controlled Waste Regulations 2012 (CWR) provides for the classification of waste (household, industrial or commercial waste), and lists the types of waste for which local authorities may make a charge for collection and disposal.

Key point

The Controlled Waste Regulations provide for the classification of waste (household, industrial or commercial), and lists the types of waste for which local authorities can charge for collection and disposal.

The Dangerous Substances and Explosive Atmospheres Regulations 2002

Dangerous substances are any substances used or present at work that could, if not properly controlled, cause harm to people as a result of a fire or explosion or corrosion of metal. Dangerous substances are used in a wide variety of engineering activities and they include such things as solvents, paints, varnishes, flammable gases, dust from machining and sanding operations, etching and cutting fluids, pressurized gases and substances corrosive to metal.

Here are some examples of the range of engineering activities covered by The Dangerous Substances and Explosive Atmospheres Regulations 2002 (DSEAR):

- storage of petrol as a fuel for vehicles and generators
- use of flammable gases, such as acetylene, for welding
- handling and storage of flammable wastes such as fuel oils
- welding or other 'hot work' on tanks and drums used with flammable material

- work that could release naturally occurring flammable substances such as methane in coal mines or at landfill sites
- use of flammable solvents in a workshop environment
- storage and display of flammable materials such as paints and lubricants
- filling, storing and handling aerosols with flammable propellants such as liquid petroleum gas (LPG)
- transporting flammable substances in containers around a workplace
- deliveries from road tankers, such as petrol and bulk powders
- chemical manufacturing, processing and warehousing
- the petrochemical industry, both onshore and offshore
- handling, storage and use of gases under pressure
- handling, storage and use of corrosive substances.

DSEAR requires employers to control the risks and protect people from exposure to dangerous substances. In particular, employers must:

- find out what dangerous substances are present in the workplace and the risks associated with them
- put control measures in place to either remove those risks or, where this is not possible, control them
- put controls in place to reduce the effects of any incidents involving dangerous substances
- prepare plans and procedures to deal with accidents, incidents and emergencies involving dangerous substances
- make sure employees are properly informed about and trained to control or deal with the risks from dangerous substances
- identify and classify areas of the workplace where explosive atmospheres may occur and avoid ignition sources in those areas.

The Chemicals (Hazard Information and Packaging for Supply) Regulations 2002

The Chemicals (Hazard Information and Packaging for Supply) Regulations 2002 (CHIP) describes the procedures for classifying dangerous substances and dangerous preparations, and describes the safety data sheets that must be provided when dangerous substances and dangerous chemical preparations are supplied.

CHIP is intended to protect people and the environment from the harmful effects of dangerous chemicals. CHIP requires that dangerous chemicals must be appropriately and safely packaged, labelled and accompanied by additional information for safe use (such as Material Safety Data Sheets). CHIP also requires that advertisements for dangerous substances and dangerous preparations must refer to any associated hazards.

Test your knowledge 4.1

Under which environmental regulations should:

a) waste material be classified as household, industrial or commercial?

b) chimneys of gas-burning installations be built to a minimum height?

c) employees be informed of the hazards associated with handling and using solvents?

d) permits be obtained for the construction and operation of a waste processing plant?

Test your knowledge 4.2

List **four** engineering activities that are covered by DSEAR.

Test your knowledge 4.3

List **four** things that DSEAR requires employers to do in order to control the risks and protect people from exposure to dangerous substances.

Learning outcome 4.2

Explain what is contained in the environmental management systems BS EN ISO 14001 (EMS)

Organizations, large and small, now recognize the need to minimize the harmful effects on the environment caused by their activities. To respond to this challenge many companies, including those active in the engineering sector, are moving towards the introduction of environmental management systems (EMS). Such systems involve adopting a systematic approach to incorporating energy and environmental goals as well as compliance with relevant legislation, regulations and directives into a company's day-to-day operation.

Having a formal (documented) system in place ensures that employees at all levels in the company are aware and that the system is applied consistently. The aim of an effective EMS can be stated quite simply. It is a system that enables an organization to

Key point

Environmental management is what an organization does to minimize the harmful effects on the environment caused by its activities.

reduce its environmental impact and at the same time increase its operating efficiency. By linking the two goals together (minimizing environmental impact and improving efficiency) companies are able to make a sound commercial as well as social case for introducing an EMS.

ISO 14000

The International Organization for Standardization (ISO) has published a set of standards that provides companies with a tool for managing, measuring, improving and communicating the environmental aspects of their operations. The standards are entirely voluntary but their adoption can bring specific benefits to companies that adopt them including improved efficiency and environmental awareness.

ISO 14001 is the heart of the series while the other standards provide guidance on specific aspects of environmental management. The ISO 14004 EMS guidance document explains how to implement ISO 14001. Objective evidence is necessary to fulfil the requirements of ISO 14004. This evidence needs to be *auditable* in order to demonstrate that the environmental management system is operating effectively.

Figure 4.3 shows the key elements of the ISO 14001 EMS model. They are as follows:

1. *Environmental policy* – establishes and communicates an organization's position and commitment as it relates to energy and the environment.

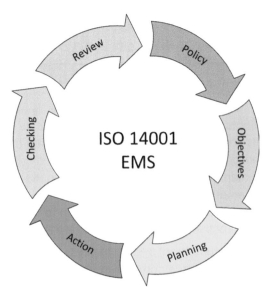

Figure 4.3 The main parts of the ISO 14001 EMS model.

2. *Planning* – identifies energy and environmental issues and requirements, and defines the initiatives and resources needed to achieve the environmental policy and economic goals.

3. *Implementation and operations* – describes the procedures, programs and responsibilities necessary to implement the key initiatives to achieve goals.

4. *Checking and corrective action* – regularly monitors and assesses the effectiveness of energy and environmental management activities.

5. *Management review* – high-level evaluation of the management system as a whole to determine its overall effectiveness in terms of driving continual improvement and achieving business goals.

Key point

Under ISO 14004 objective evidence is necessary to demonstrate that an environmental management system is operating effectively and conforming with the ISO standard.

Test your knowledge 4.4

Explain the main features and advantages of an environmental management system.

Learning outcome 4.3

Explain what climate change levy (CCL) is and its implications, and what is exempt

There are several causes of climate change. Some are natural, such as changes in ocean circulation, whilst others are undoubtedly caused by human activity, notably industrialization and the pollution of the atmosphere by greenhouse gases. By increasing the amount of greenhouse gases in the atmosphere through the burning of fossil fuels and deforestation, man has contributed to the Earth's natural greenhouse effect, resulting in a worrying increase in global average temperature.

During the last century, the average surface temperature of the Earth (about 15°C) has increased by well over 0.5°C. Much of this rise in temperature has occurred since 1980.

In the last 135 years (when reliable records have been kept) the ten warmest years, with only one exception, have all have occurred since 2000. This worrying trend is set to continue as global average temperature continues to rise.

It has been suggested by reputable authorities that reaching or exceeding a threshold of 2°C above pre-industrial levels would have a major impact on our planet's bio-systems and so all governments have had to consider ways of addressing climate change. The latest estimates made by NASA suggest that the Earth is warming at an average of 0.87°C per annum and that this rate has increased from zero in the period between 1940 and 1960. If nothing is done, the worrying 2°C threshold could well be exceeded before the end of this decade.

Activity 4.1

Visit the climate projections website of the UK Met Office by going to:

http://ukclimateprojections.metoffice.gov.uk. Search for the document, *UK Climate Projections: Briefing report* and use the report to answer each of the following questions:

1. What has been the average central England temperature rise since the 1970s?
2. What is the most likely cause of this temperature rise?
3. What is the average rate of sea level increase in the UK during the 20th century?
4. By 2080 (and relative to a 1961–1990 baseline) what is the expected range of increase of summer temperature in southern England?
5. What changes might be expected in the Atlantic Ocean circulation (Gulf Stream) in the next century?

Important note: The document relates to UK Climate Projections made in 2009 (UKCP09). This document is now out of date and, following the historic *Paris Agreement on Climate Change* in December 2015, there is an urgent need for a new set of climate projections. Unfortunately, the next set of official UK projections are not due until 2018 (UKCP18). However, if an updated report is available, you should base your answers on these rather than the 2009 projections.

The climate change levy

Several countries, including the UK, have introduced measures that are intended to reduce energy consumption by industrial and commercial users. In the UK the climate change levy (CCL) is a tax imposed on the use of energy and it is designed to encourage businesses to operate in a more environmentally friendly way.

The CCL is paid at one of two rates; the *main rate* (paid on electricity, gas and solid fuels such as coal, lignite, coke and petroleum coke) and the *carbon price support* (CPS) rate (paid for gas, liquid petroleum gas, coal and other solid fuels). The main rate is paid by industry, commerce, agriculture and public services whilst the CPS rate is paid by operators of electricity generating stations and operators of combined heat and power (CHP) stations.

Exemptions to the climate change levy

There are several exemptions to the CCL including:

- fuel used by domestic or transport sectors
- fuel used for the production of other forms of energy
- very small firms using a domestic amount of energy (e.g. not exceeding 1000 kW hours per month of electricity)
- fuel oils which is already subject to excise duty
- electricity generated from renewable energy
- fuels used jointly as a feedstock or as an energy source within the same process (e.g. coke in steel-making)
- electricity used in electrolysis processes (e.g. the chlor-alkali process and primary aluminium smelting).

The enhanced capital allowance scheme

In an initiative to improve energy efficiency, the UK Government has introduced an incentive for companies to invest in energy-efficient capital plant and equipment. The ETL (or Energy Technology Product List) is a government-managed list of energy-efficient plant and machinery. Equipment on this list qualifies for the enhanced capital allowance tax scheme for UK businesses. The list includes products such as:

- air-to-air energy recovery systems
- automatic monitoring and targeting equipment
- boiler equipment
- combined heat and power
- compressed air equipment
- heat pumps
- heating, ventilation and air-conditioning equipment
- high speed hand air dryers
- lighting
- motors and drives
- pipework insulation
- refrigeration equipment
- solar thermal systems
- uninterruptible power supplies
- warm air and radiant heaters
- waste heat to electricity conversion equipment.

Note that, for a product to be on the ETL, it must meet specific energy-saving or energy-efficient criteria. The Department for Business, Energy and Industrial Strategy (DBEIS) annually reviews the technologies and products that qualify for inclusion. The ETL is managed on behalf of DBEIS by the Carbon Trust. Provided that it

Key point

The climate change levy (CCL) is a tax imposed on the use of energy and it is designed to encourage businesses to operate in a more environmentally friendly way.

is included on the ETL at the time of purchase, a business that pays income or corporation tax is able to claim 100% first year capital allowance on the product.

Key point

The Energy Technology Product List (ETL) is a list of government-approved energy-efficient plant and equipment. Products included on this list qualify for enhanced capital allowance.

Test your knowledge 4.5

Explain how an engineering company might be able to claim enhanced capital allowance for the purchase of an air-conditioning plant that is energy-efficient.

Test your knowledge 4.6

List **four** different items of plant and equipment likely to be found on the Energy Technology Product List (ETL). What are the criteria for a product to be included in the list?

Test your knowledge 4.7

List **four** exemptions to the climate change levy (CCL).

Learning outcome 4.4

Describe what sources of energy are available other than fossil fuels

In order to reduce the effects of climate change and cope with dwindling supplies of fossil fuels there is now an urgent need to move to energy sources that are both cleaner and renewable. Various sources of energy are currently being explored. Each has its own particular advantages and disadvantages. The most important point, however, is that these alternative energy sources all have significantly lower carbon emissions that would otherwise contribute to greenhouse gases and global warming. Here is a list of some of the potential energy sources:

- solar energy
- heat pump
- hydroelectric (reservoirs)
- wave power (tidal)
- wind power (wind farms)
- waste end energy-producing incineration.

Table 4.1 provides more information on the main alternatives to fossil fuel energy.

Solar energy

From an environmental perspective photovoltaic (PV) energy or *solar power* offers huge benefits. In recent years the cost of installing PV systems has fallen significantly but the lifetime cost can still be rather high, particularly when taking into account the amount of land required for a solar farm. Another more obvious problem associated with solar energy is that, although the supply is infinite, the amount of energy produced varies with the amount of sunlight and no energy at all is produced during the hours of darkness.

Figure 4.4 A solar farm.

Wind energy

There is nothing new about wind energy. Man has been using it for thousands of years, both as a means of propulsion and as a source of energy for grinding grain and pumping water. Today, and with generating costs falling with increasing use, people are beginning to realize that wind power is one of the most promising energy sources. The principal disadvantage of wind energy is that it is not a constant course of energy and many sites are unsuitable due to lack of wind.

In addition to the energy sources mentioned in Table 4.1 there are several other potential sources of energy including biomass and geothermal.

Key point

Alternative energy sources have significantly lower carbon emissions when compared with conventional fossil fuels.

Figure 4.5 An offshore wind farm.

Table 4.1 Alternative sources of energy.

Energy source	Main aspects	Advantages	Disadvantages
Solar	Photovoltaic (PV) cells generate electricity from sunlight.	Potentially infinite supply of energy. Requires no fuel. Cost of PV installations has been falling significantly. Minimal maintenance required. Electrical energy generated can be sent over long distances.	Requires sunlight (non-constant output). High initial cost. A large area of land is needed for a commercial installation.
Heat pumps	Ground source heat pumps use buried pipes to extract heat from the ground. Air source heat pumps extract heat from the air. With both types, heat is absorbed into a circulating fluid and then passed into a heat exchanger.	Potentially infinite supply of energy. Requires no fuel. Needs only minimal maintenance.	GSHP requires a large area of land (or a deep borehole). Only suitable for local heating and hot water supplies.

(Continued)

Table 4.1 *(Continued)*

Energy source	Main aspects	Advantages	Disadvantages
Hydroelectric	Water-driven turbines generate electricity. Supplies of water can be taken from free-flowing rivers or from lakes and reservoirs via a dam.	Capable of generating large amounts of energy at low lifetime cost. Electrical energy generated can be sent over long distances.	Requires large amounts of water. Relatively high initial cost of construction. May have a significant impact on local hydrology.
Wave power	The movement of tides drives turbines located on barrages or on the sea bed. Tidal barrages can be placed across estuaries. Future installations may be built offshore and use sea currents or wave motion rather than tides.	Ideal for estuaries and coastal areas. Electrical energy generated can be sent over long distances.	Relatively high initial cost of construction. High cost of maintenance offshore. Many estuaries may be unsuitable.
Wind power	Wind turbines generate electricity. Wind farms can be located onshore or offshore where winds tend to be stronger and more constant.	Relatively high output. Electrical energy generated can be sent over long distances.	Requires wind (non-constant output). Not suitable for all geographic locations. Some people consider them unsightly. High cost of maintenance offshore.
Waste incineration	Combustible waste materials are incinerated to produce hot water and steam. Steam can be used to drive turbines and produce electricity.	Relatively inexpensive and provides a means of waste disposal. Relatively low lifetime cost.	Not all waste products are suitable for incineration. Harmful gases and residues may be produced. Atmospheric pollutants need to be removed from combustion products.

Biomass

Biomass is a renewable energy source derived from organic matter such as plant and animal waste. Many biomass power stations use wood residue (such as bark and sawdust) to generate electricity. There are several ways to produce energy from biomass including burning biomass to generate heat and supply steam turbines that produce electricity, turning feedstocks into liquid biofuels, and harvesting gas from landfill or anaerobic digesters.

Figure 4.6 A biomass energy plant.

Geothermal

Geothermal uses the Earth's own heat to generate electricity using steam turbines. This source of energy is more appropriate in areas where there is a significant amount of geothermal activity (such as Iceland and New Zealand) and it usually involves pumping cold water underground and recovering steam where it emerges. Geothermal energy is capable of generating large amounts of energy at relatively low cost but it can be unpredictable and installations need to be carefully designed and built for a specific location.

Figure 4.7 A geothermal energy plant.

Test your knowledge 4.8

List **two** advantages and **two** disadvantages of each of the following energy sources:

a) solar (PV)

b) wind

c) tidal

Test your knowledge 4.9

An engineering company located in a remote moorland area wishes to explore the use of alternative energy sources to provide power for heating and lighting. Which **two** alternative energy sources might be considered suitable and why?

Learning outcome 4.5

Evaluate the criteria with regard to emissions

Modern engineering processes and systems are increasingly designed and implemented to minimize environmental impact. You need to be able to identify how individual engineering companies seek to minimize the impact of their activities on the environment by:

- designing plant and products that optimize energy use and minimize pollution
- employing good practice such as the efficient use of resources, recycling and the use of techniques to improve air and water quality
- using management review and corrective action (e.g. through the use of an EMS)
- ensuring compliance with relevant legislation and regulations.

Many engineering activities involve the processing of materials. Some of these materials occur naturally and, after extraction from the ground, may require only minimal treatment before being used for some engineering purpose. Examples are timber, copper, iron, silicon, water and air. Other engineering materials need to be manufactured. Examples are steel, brass, plastic, glass, semiconductors such as gallium arsenide, and ceramic materials. The processing and use of these materials can produce effects which may have an impact on the environment.

Economic and social effects

Economic and social effects stem from the wealth generated by the extraction of raw material and its subsequent processing or manufacture into useful engineering materials. For example, the extraction of iron-ore in Cleveland and its processing into pure iron and steel has brought great benefit to the Middlesbrough region. The work has attracted people to live in the area and the money they earn tends to be spent locally. This benefits trade at the local shops and entertainment centres, and local builders must provide more homes and schools, and so on. The increasing population produces a growth in local services which includes a wider choice of different amenities, better roads and communications and arguably, in general, a better quality of life.

Effects on the landscape

The extraction of raw materials can make the landscape ugly and untidy. Heaps of slag around coal mines and steelworks together with holes left by disused quarries are not a pretty sight. In recent years much thought and effort has been expended on improving these eyesores. Slag heaps have been remodelled to become part of golf courses and disused quarries filled with water to become centres for water sports or fishing. Disused mines and quarries can also be used for taking engineering waste in what is known as a *landfill* operation prior to the appropriate landscaping being undertaken. Other potential problems can arise from having to transport the raw materials used in engineering processes from place to place. This can have an adverse effect on the environment resulting from noise and pollution.

Pollutants

Engineering activities are a major source of *pollutants* that in turn can result in *pollution*. Air, soil, rivers, lakes and seas are all, somewhere or other, polluted by waste gases, liquids and solids discarded by the engineering industry. Because engineering enterprises tend to be concentrated in and around towns and other built up areas, they tend to be common sources of pollutants.

Electricity is a common source of energy and its generation often involves the burning of the *fossil fuels*: coal, oil and natural gas. In so doing, each year, billions of tonnes of carbon dioxide, sulphur dioxide, smoke and toxic metals are released into the air to be distributed by the wind. The release of hot gases and hot liquids also produces another pollutant; heat. Some electricity generating stations use nuclear fuel which produces a highly radioactive solid waste rather than gases.

The generation of electricity is by no means the only source of toxic or biologically damaging pollutants. The exhaust gases from motor vehicles, oil refineries, chemical works and industrial furnaces are other problem areas. Also, not all pollutants are graded as *toxic*. For example, plastic and metal scrap dumped on waste tips, slag heaps around mining operations, old quarries, pits and derelict land are all *non-toxic*. Finally, pollutants can be further defined as *degradable* or *non-degradable*. These terms simply indicate whether the pollutant decomposes or disperses itself with time. For example, smoke is degradable but dumped plastic waste is not.

Figure 4.8 Atmospheric pollution caused by engineering activities.

Carbon dioxide

Carbon dioxide in the air absorbs some of the long-wave radiation emitted by the Earth's surface and in so doing is heated. The more carbon dioxide there is in the air, the greater the heating or greenhouse effect. This is a major cause of global warming, resulting in a global average temperature rise, as mentioned earlier in Section 4.3. In addition to causing undesirable heating effects, the increased quantity of carbon dioxide in the air, especially around large cities, may lead to people developing respiratory problems.

Oxides of nitrogen

Oxides of nitrogen are produced in most exhaust gases, and nitric oxide is prevalent near industrial furnaces. Fortunately, most oxides of nitrogen are soon washed out of the air by rain. But if there is no rain, the air becomes increasingly polluted and unpleasant.

Sulphur dioxide

Sulphur dioxide is produced by the burning of fuels that contain sulphur. Coal is perhaps the major culprit in this respect. High concentrations of this gas cause the air tubes in people's lungs to constrict and breathing becomes increasingly difficult. Sulphur dioxide also combines with rain droplets eventually to form sulphuric acid or *acid rain*. This is carried by the winds and can fall many hundreds of miles from the sulphur dioxide source. Acid rain deposits increase the normal weathering effect on buildings and soil, corrode metals and textiles, and damage trees and other vegetation.

Smoke

Smoke is caused by the incomplete burning of the fossil fuels. It is a health hazard on its own but even more dangerous if combined with fog. This poisonous combination, called *smog*, was prevalent in the early 1950s, and formed in its highest concentrations around the large cities where many domestic coal fires were then in use. Many deaths were recorded, especially among the elderly and those with respiratory diseases. This led to the first Clean Air Act which prohibited the burning of fuels that caused smoke in areas of high population. So-called *smokeless zones* were established.

Dust and grit

Dust and grit (or *ash*) are very fine particles of solid material that are formed by combustion and other industrial processes. These are released into the atmosphere where they are dispersed by the wind before falling to the ground. The lighter particles may be held in the air for many hours. They form a mist, which produces a weak, hazy sunshine and less light.

Toxic metals

Toxic metals, such as lead and mercury are released into the air by some engineering processes and especially by motor vehicle exhaust gases. The lead and mercury can be carried over hundreds of miles before falling in rainwater to contaminate the soil and the vegetation it grows. Motor vehicles now use unleaded petrol and this has led to a reduction in the lead pollution that was previously a problem when leaded fuel was used.

Ozone

Ozone is a gas that exists naturally in the upper layers of the Earth's atmosphere. At that altitude it is one of the Earth's great protectors but should it occur at ground level it is linked to pollution. *Stratospheric ozone* shields us from some of the potentially harmful excessive ultraviolet radiation from the sun. In the 1980s it was discovered that emissions of gases from engineering and other activities were causing a 'hole' in the ozone layer. There is concern that this will increase the risk of skin cancer, eye cataracts, and damage to crops and marine life.

At ground level, sunlight reacts with motor vehicle exhaust gases to produce ozone. Human lungs cannot easily extract oxygen (O_2) from ozone (O_3) so causing breathing difficulties and irritation to the respiratory channels. It can also damage plants. This ground level or *tropospheric ozone* is a key constituent of what is called *photochemical smog* or *summer smog*. In the UK it has increased by about 60% in the last 40 years.

Heat

Heat is a waste product of many engineering activities. A typical example being the dumping of hot coolant water from electricity generating stations into rivers or the sea. This is not so prevalent today as increasingly stringent energy-saving measures are applied. However, where it does happen, river and sea temperatures can be raised sufficiently in the region of the heat outlet to destroy natural aquatic life. As a consequence, *energy efficiency* and the reduction of waste heat is an increasingly important consideration in many engineering processes.

Chemical waste

Chemical waste dumped directly into rivers and the sea, or onto land near water, can cause serious pollution which can wipe out aquatic life in affected areas. There is also the long-term danger that chemicals dumped on soil will soak through the soil into the ground water which we use for drinking purposes and which will therefore require additional purification.

Radioactive waste

Radioactive waste from nuclear power stations or other engineering activities which use radioactive materials poses particular problems. Not only is it extremely dangerous to people – a powerful cause of cancer – but its effects do not degrade rapidly with time and remain dangerous for scores of years. Present methods of disposing of radioactive waste, often very contentious, include their encasement in lead and burial underground or at sea.

Noise and vibration

Noise and vibration is another emission associated with engineering activities. With larger, heavier plant this can be a particular problem. In order to reduce noise and vibration careful consideration needs to be given to the design and construction of engineering plant and buildings. For example, noise can be reduced by muffling exhausts and chimneys while vibration can be reduced by installing engines and plant on mountings that prevent the transmission of vibration to floors and surrounding walls.

Figure 4.9 Noise and vibration can often be a problem with heavy plant and equipment.

Figure 4.10 Diesel generator with muffled exhaust to reduce noise.

Derelict land

Derelict land is an unfortunate consequence of some engineering activities. Derelict land is land so badly damaged that it cannot be used for other purposes without further treatment and it includes land that is disused or abandoned land requiring extensive restoration work to bring it into use or improve its appearance. Land is often made derelict by mining and quarrying operations, the dumping of waste or by disused factories from abandoned engineering activities.

Test your knowledge 4.10

List **three** different types of air pollution resulting from engineering activities.

Test your knowledge 4.11

Explain how damage to the ozone layer has resulted in an increase in the incidence of skin cancer.

Activity 4.2

Investigate the effects on the physical environment to include *human*, *natural* and *built* of **one** engineering activity selected from **each** of the following categories: *production*, *maintenance* and *materials handling*. Write a brief report based on your investigations. Your report should include a brief description of:

- the environmental effects of the materials used
- the short-term and long-term environmental effects of any waste products
- relevant environmental legislation effects giving specific examples.

Examples of engineering production activities are:

- motor car manufacture
- steel manufacture
- coal mining.

Examples of engineering maintenance activities are:

- car dealerships and garages
- local council road maintenance depots
- maintenance of electricity and gas supplies.

Examples of materials handling activities are:

- container handling terminals
- moving cargo by rail and road
- conveying goods on moving belts.

Hints:

1. Make sure that your selected engineering activity gives you the opportunity to produce the necessary amount of evidence to demonstrate your competence and understanding.
2. You could approach this activity through case studies (e.g. those involving court action concerning failure to comply with legislation).
3. It is important to be clear about the difference between *waste products* and *by-products*. The by-products from one process can be sold as the raw materials for other processes. For example, natural gas is a by-product of oil extraction and a useful fuel used in the generation of electricity. Waste products are those that cannot be sold and may attract costs in their disposal. Nuclear power station waste is a typical example.

Learning outcome 4.6

Recognize the requirements for safe disposal of waste

All organizations must comply with relevant legislation when disposing of waste. This is particularly important when the waste can be classified as hazardous. However, before deciding to dispose of waste it is good practice to consider the following factors:

1. *Prevent* – could the company avoid using the material?
2. *Minimize* – could the company use fewer and/or less hazardous materials?
3. *Re-use* – could the material or substances be re-used?
4. *Recycle* – could the material or substance be recycled?
5. *Recover* – could active materials or substances be recovered or could energy be generated from incinerating the waste?
6. *Disposal* – the material or substance could be sent for landfill or incineration without energy recovery.

These six factors are often referred to as the *waste hierarchy* (see Figure 4.11). This highlights the need to extract the maximum practical benefits from products whilst, at the same time, minimizing the amount of waste. By applying the waste hierarchy, it is possible to reduce the emission of greenhouse gases, minimize pollutants, save energy and conserve limited resources of fossil fuels.

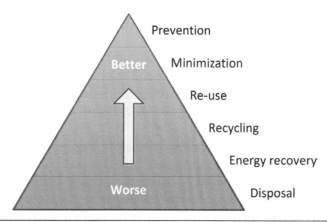

Figure 4.11 The waste hierarchy.

Duty of care

The Environmental Protection (Duty of Care) Regulations 1991 place obligations on those producing waste to ensure that any waste they produce is handled safely and in accordance with the law. Most hazardous waste needs a very accurate description, so, under the duty of care, engineering companies must provide sufficient description to enable their waste to be safely managed. Waste must always be transferred to an authorized person, e.g. a registered carrier or licensed waste manager. Records of assignments must be kept for three years. If waste is moved by road it may also be subject to the requirements for the transport of dangerous goods.

Hazardous waste

Waste is generally considered hazardous if it (or the material or substances it contains) is potentially harmful to humans or the environment. Examples of hazardous waste encountered in engineering include:

- asbestos
- chemicals (e.g. brake fluid)
- batteries
- solvents
- oils
- refrigerants (and other substances causing ozone depletion)
- hazardous waste containers.

Waste disposal

Where waste is produced, stored or removed from an engineering company's premises the following steps should be followed:

1. Classify the waste materials and check if they should be treated as hazardous.
2. Separate and store hazardous waste safely (including clear marking).
3. Use an authorized business to collect, recycle or dispose of the waste (also check that waste carriers and waste disposal sites have relevant environmental permits).
4. Complete a *waste transfer note*. These need to provide a description of the waste and how it is contained (e.g. skip, drum, etc.), details of the current holder of the waste (the *transferor*), the person collecting the waste (the *transferee*), and the details of the transfer or collection point. Copies of waste transfer notes must be retained for inspection and future reference.
5. Maintain a record (known as a *register*) for three years at the premises where the waste is produced or stored.

Key point

Engineering companies must keep accurate records of all consignments of hazardous waste.

Test your knowledge 4.12

List **six** different engineering waste materials and substances that are considered hazardous.

Test your knowledge 4.13

Which of the following waste materials should be considered hazardous and why?

1. Offcuts of aluminium sheet.
2. Chemically contaminated PPE.
3. Brake fluid.
4. Copper-clad cable.
5. Metal cutting swarf removed from a machine tool.
6. Unpopulated printed circuit boards.
7. Fluorescent lights.
8. Machine cutting oil.

Activity 4.3

Find out what happens to the general waste such as coolant, lubricants, swarf and rags, used in your workshop. Classify the various different types of waste and describe how each type of waste is handled and disposed.

Activity 4.4

A local firm, Avon Haulage, wishes to provide a new service to customers involving the collection and transport of hazardous waste. Visit the UK Government website, https://www.gov.uk, obtain information of the Hazardous Waste Regulations and use it to determine the steps and procedures that the company must follow in order to comply with UK regulations concerning the transportation of hazardous waste.

Accidental spillage, leakage and release of substances

Since the inadvertent release of some substances can have serious consequences, it is essential for engineering companies to have plans in place to minimize the likelihood of spills and leakages and also to deal with them effectively if and when they do occur. Effective training and reporting systems need to be implemented and personnel need to be familiar with the correct procedure for dealing with substances that are spilled or leaked. Typical examples might be fuel leaking from corroded pipework, ruptured storage tanks, or over-filling containers used for transportation. First and foremost, it is essential to know what the substance is and how it should be dealt with. Consequently, clear and consistent labelling is essential.

Figure 4.12 Safe storage of chemicals.

Accidental release measures are usually included in Material Safety Data Sheets (MSDS). These cover personal and environmental precautions necessary in the case of spills and accidental leakages. Accidental release measures should describe the dangers related to the substance or product. Special attention should be paid to factors that are not obvious at first sight, like danger of slipping, ignition of combustible air–gas mixtures which spread on the floor etc. Instructions for cleaning up or picking up spilt product should also be provided. Figure 4.13 shows an extract relating to the handing of a spill or leakage of crude oil.

Ebsworth Plant Accidental Release Measures

1. NOTIFICATION
In the event of a spill or accidental release, notify Head Office and relevant authorities in accordance with all applicable regulations.

2. PROTECTIVE MEASURES
Avoid contact with spilled material.
Warn or evacuate occupants in surrounding and downwind areas if required due to toxicity or flammability of the material.
See also: Fire Fighting Information, First Aid Advice and Personal Protective Equipment.

3. LAND SPILL MANAGEMENT
Eliminate all ignition sources (no smoking, flares, sparks or flames in immediate area).
Stop leak if you can do it without risk.
All equipment used when handling the product must be grounded.
Do not touch or walk through spilled material.
Prevent entry into waterways, sewer, basements or confined areas.
Use a vapor suppressing foam may be used to reduce vapours.
Note: Water spray may reduce vapour but may not prevent ignition in closed spaces.
Absorb or cover with dry earth, sand or other non-combustible material and transfer to containers.

4. WATER SPILL MANAGEMENT
Stop leak if you can do it without risk.
Remove from the surface by skimming or with suitable absorbents.
If permitted by regulatory authorities the use of suitable dispersants should be considered where indicated in local oil spill contingency plans.

5. ENVIRONMENTAL PRECAUTIONS
Use booms as a barrier to protect the shorelime.
Use containment booms when the ambient temperature is below the flash point of the material.
Prevent entry into waterways, sewers, basements or confined areas.

Figure 4.13 Accidental release measures for crude oil petroleum.

Key point

Material Safety Data Sheets (MSDS) usually include accidental release measures that provide details of the environmental precautions necessary in the case of spills and accidental leakages.

Finally, on sites where hazardous materials are produced, stored or used, an effective spill prevention and response plan should:

- identify all potential spill areas
- specify material handling procedures
- describe response procedures
- provide appropriate clean-up equipment.

Pollution prevention and health and safety should be paramount in any spill response plan.

Activity 4.5

Refer to the accidental release measures shown in Figure 4.13 and use them to answer the following questions:

1. List **three** potential ignition sources.
2. What can be used to reduce vapours?
3. How can the product be removed from the surface of a stretch of water?
4. How can the product be removed if it is spilled on land?
5. How can shorelines be protected from contamination?
6. What special precaution should be taken when using equipment and handling the product?
7. Why is it necessary to avoid the product entering basements and confined areas?

Test your knowledge 4.14

List **four** key elements of a plan to cope with the accidental spillage of a hazardous waste material.

Review questions

1. Describe the basic content and application of the Environmental Protection Act.

2. Explain the main features of:
   a) the Pollution Prevention and Control Act
   b) the Controlled Waste Regulations
   c) the Clean Air Act

3. Describe the basic content and application of an environmental management system (EMS).

4. What impact is an EMS likely to have on the day-to-day operation of an engineering company? Explain your answer.

5. List **four** different types of hazardous waste likely to be produced by a typical engineering company.

6. List the **six** factors that constitute the 'waste hierarchy'.

7. What is an 'alternative source of energy' and how does it differ from a conventional fossil fuel energy source?

8. Describe **four** different alternative energy sources. Give **one** advantage and **one** disadvantage of each of these sources.

9. Compare and contrast the environmental impact of each of the following energy sources:
   a) tidal
   b) wind
   c) waste incineration

10. Explain the correct procedure for the safe disposal of hazardous waste such as cutting oil and swarf.

Chapter checklist

Learning outcome	Page number
4.1 Analyse the relevant legislation and EU directives with regard to environmental management	82
4.2 Explain what is contained in the environmental management systems BS EN ISO 14001	87
4.3 Explain what climate change levy (CCL) is and its implications, and what is exempt	89
4.4 Describe what sources of energy are available other than fossil fuels	92
4.5 Evaluate the criteria with regard to emissions	97
4.6 Recognize the requirements for safe disposal of waste	104

Engineering organizational efficiency and improvement

CHAPTER 5

Understanding production activities

Learning outcomes

When you have completed this chapter you should understand production activities, including being able to:

5.1 Explain the different types and methods of production.

5.2 Recognize the considerations that need to be made when selecting a production type or method.

5.3 Identify the different stages of production planning.

5.4 Explain how to apply typical process charts.

Chapter summary

Manufacturing production is the core activity performed by many engineering companies. Production can appear deceptively simple but, in reality, it is extremely complex. Because of this we need to find ways of understanding the process so that it can be effectively managed and controlled. This is all part and parcel of the everyday work of the production engineer and this chapter will provide you with an insight into what this important work involves.

Learning outcome 5.1

Explain the different types and methods of production

Production is at the heart of any manufacturing business. The production process translates the designs for products, which are based on market analysis, into the goods wanted by customers. Depending on the product type and the quantities involved, production can take various forms including mass, continuous flow, intermittent flow, batch and one-off 'job shop' manufacture. You need to know how these methods differ and in what situations they are used. However, whichever method of production is chosen, the ultimate goal is always to ensure that a product is produced in the most efficient and cost-effective way.

Mass production

Mass production generally refers to the manufacture of very large quantities of identical products. This is usually achieved using a continuous flow process where the various stages of production follow seamlessly from one to the next. Mass production is appropriate whenever the market for a particular product is very large and is expected to remain large in the conceivable future. For example, the manufacture of LED lamps for the domestic market in the UK is both sizeable and projected to increase over the next few years, making high-volume mass production appropriate. Mass production is invariably based on continuous or intermittent flow.

Flow production

Flow production involves a linear sequence of operations with each stage in the process leading to the next. For example, the manufacture of a washing machine might involve the manufacture of a stainless steel drum which is inserted into a stainless steel outer casing before being attached to an electric motor via a belt drive. The various stages in the manufacturing process must be performed in the correct sequence.

Continuous flow production

Continuous production describes a manufacturing process where a product moves through a series of processes without stopping. This type of production is only appropriate for very high-volume manufacturing and it usually involves a high degree of automation which, in turn, demands significant investment in plant and equipment.

A particular feature of continuous flow production is that each stage of the process must be designed so that it can be completed within a similar time frame. Failure to observe this requirement can result in delays and inefficient use of resources.

Intermittent flow production

Intermittent flow relates to a manufacturing process where the flow is temporarily interrupted after one or more stages of manufacturing. The flow is then restarted at some later time so that the remaining processes can be applied.

Note that things can change when a manufacturing company expands the scale of its operation and, as a result, needs to change its production methods. Typical examples might be to change production method from batch to flow, or from intermittent flow to continuous flow. These changes can be expected to increase production capacity. However, other constraints, such as lack of capital for investment in new plant and equipment, may also impact on production capacity.

Job shop production

This term, which might at first sound a little old-fashioned, refers to a type of manufacturing where products are individually made. Job shop production is sometimes also referred to as one-off or project-based production and all three terms are applied to methods of production in which one product is completed before the next one is started. With job shop production each manufacturing project is unique and the manufactured products are invariably customized in some way so that no two products are identical.

A significant advantage of job shop production is that it allows for a high degree of customization. Because of this, manufacturers can

Key point

Flow production involves a continuous movement of products through a series of manufacturing processes. When one process is complete the next will start, and so on. To be efficient (i.e. to ensure that all of the manufacturing equipment is being used all of the time) this type of production requires that individual processes be accomplished within a similar time frame. Flow production is appropriate when a large number of products are required on a continuous basis.

Figure 5.1 A typical job shop where products can be customized to a high degree.

charge premium prices for their products but the downside is that, due the high cost, the market may be rather limited. However, job shop production is inherently flexible and a wide range of different processes are usually available. A significant disadvantage of job shop production is that resources such as machine tools and other manufacturing equipment may only be used on an intermittent basis and, for that reason, it may be difficult to justify the initial capital investment. Staff with specialized skills may also be under-utilized and, because of this, those working in a job shop environment should ideally be multi-skilled and be familiar with a wide range of different processes.

Test your knowledge 5.1

1. Explain why continuous flow production is less flexible than job shop production.
2. State one advantage and one disadvantage of job shop production.
3. Give **two** examples of products that are suitable for flow production.
4. Give **two** examples of products that are suitable for job shop production.

Recognize the considerations that need to be made when selecting a production type or method

Selecting the right type of production process is vitally important and is a strategic decision taken at management level in most businesses. The chosen method of production will commit a company to particular kinds of equipment and labour force because the large capital investments that have to be made will undoubtedly limit future options. For example, a car manufacturer has to commit very large expenditure to lay down plant for production lines to mass produce cars. Once in production the company is committed to the technology and the capacity created for a long time into the future.

In addition to the initial capital investment there are a number of other factors that need to be taken into account when choosing a production type or method. These factors include:

- market requirements
- product design
- availability of plant and equipment
- layout of plant and equipment
- available personnel
- production control
- quality control
- production and manufacturing cost.

We will examine each one of these considerations separately.

Market requirements

Market needs and requirements must be understood at a very early stage. Particular questions that need to be answered include an accurate forecast of demand and also the pattern of demand over the predicted life of the product. For example, a company that manufactures life rafts might predict an initial demand of 1,000 units needing to be fulfilled in the first year of manufacture followed by 250 units per annum for the next four years. Significant resources might need to be put into delivering the first-year target of 1,000 units but, thereafter, less resources will be required and production equipment could then be made available for another similar product such as an inflatable boat for the leisure market.

Product design

Another important consideration is the design of the product. If the product consists of a limited number of components requiring relatively straightforward manufacturing processes (such as pressing, stamping or moulding) the production method will be less complex than if more complex processes (such as turning, welding and soldering) are required. This serves to underline the importance of effective liaison between design and production engineering personnel at an early stage in the product design process. After all, there's little point in designing a product that is prohibitively expensive to manufacture or that might require expensive one-off capital investment. For this reason, most manufacturing companies take a very cautious approach to new product development and most seek to use existing, tried and tested, manufacturing methods.

In order to simplify the manufacturing process there is a need for product designers to:

* reduce the number of parts used
* use only standardized parts
* develop modular designs
* design parts so that they can be multi-functional
* ensure straightforward manufacturing and assembly
* minimize handling.

Case study: LED lamps

In 2013, the UK LED lighting market was estimated to be worth £330 million at manufacturers' selling prices. This market includes all finished, mains-operated LED lighting products for both the domestic and non-domestic market. The market is currently experiencing significant growth as this new technology becomes more widely accepted and it has been further encouraged by energy cost savings and concerns about the environmental impact of continued use of conventional filament lamps. Hitherto, the market is largely reliant on total system upgrades and new installation applications but as low-energy LED light bulbs become increasingly attractive as a replacement for older, inefficient filament lamps the global market is expected to grow rapidly, with an estimated market value in excess of £1,050 million by 2018.

Production costs have fallen progressively as manufacturing processes have improved. Manufacturers may choose to use a cluster of low-power LEDs which have a low light output individually, but when grouped together produce light levels

Figure 5.2 LED lamps.

sufficient for domestic lighting. Others may choose to use a small number of jumbo LEDs or multi-chip arrays, each capable of emitting significantly more light than a single low-power LED.

In 2007 a typical first-generation 650 lumen LED lamp would have a single unit retail cost of around £100 and would have used 42 individual LEDs soldered by hand. The electrical power required was about 12.5 W but the light output is the same as that of a conventional tungsten filament lamp rated at 40 W. By 2011 the equivalent low-energy lamp used only five LEDs to deliver the same 650 lumens but from an electrical power of only 9 W (showing an increase in efficiency). Instead of soldering, the individual parts were assembled using push-fitting clips (no soldering required) and the one-off retail cost had fallen to around £15. Today's 'Next Generation' 650 lumen LED bulb operates from a power of a mere 7 W and can make use of a single LED array (reducing the need for multiple internal electrical connections). The retail price of these has fallen below £4 for one-off quantities and considerably less when purchasing in bulk.

The main components of an LED lamp are shown in Figure 5.3. The process of manufacturing can be broken down into a number of stages from substrate production, LED die fabrication and packaged LED assembly (see Figure 5.4).

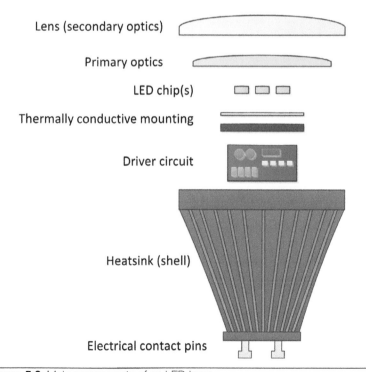

Lens (secondary optics)

Primary optics

LED chip(s)

Thermally conductive mounting

Driver circuit

Heatsink (shell)

Electrical contact pins

Figure 5.3 Main components of an LED lamp.

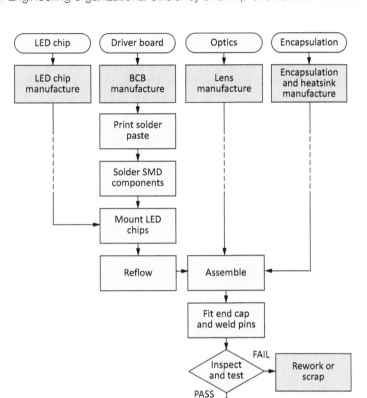

Figure 5.4 Manufacturing stages for an LED lamp.

LED manufacturing is a complex and highly technical process and is only cost-effective when large-scale production is involved. The main component parts of an LED lamp are:

- *LED chip*. The LED chip produces light output when fed from a low-voltage DC source. The colour of the light output depends on the semiconductor material used for the LED chip.
- *Printed circuit board (PCB)*. The driver circuit and LED chip are mounted on the PCB. The copper tracks on the surface of the PCB also provide a means of conducting heat away from the LED chip.
- *Driver circuit*. The driver circuit steps down and converts the 230 V AC supply to the low-voltage DC required by the LED chip.
- *Lens*. Made from PMMA material, the lens or 'secondary optic', is used to focus the light emitted by the LED chip. This makes it is possible to create a narrow beam for highly focussed light, to illuminate a painting for instance, or a wider angle for general illumination.
- *Heatsink*. LED performance depends upon temperature. Thus, even though LEDs generate very much less heat than filament

lamps, the generated heat must still be dissipated. Due to its good thermal conductivity, die-cast aluminium alloy is a common choice for the heatsink of an LED lamp.

Test your knowledge 5.2

1. Explain why continuous flow mass production methods are used for the manufacture of LED lamps.
2. Give **two** reasons why energy efficiency is important in the design of LED lamps.
3. In the design of LED lamps, list **three** improvements that have been instrumental in reducing costs.

Availability of plant and equipment

Careful consideration needs to be given to the available plant and equipment. For example, if plant and equipment is already committed to the manufacture of other products how much of it could be made available for the manufacture of new products? Alternatively, if new plant and equipment has to be purchased, how much will it cost and what will its expected life be?

Layout of plant and equipment

The layout of plant and equipment is also important, particularly when there are significant differences in the design and nature of the product being manufactured. For example, the layout of a press shop used to form small vehicle bodywork components would be unsuitable for forming long sections of steel trunking used for electrical cables. Similarly, the layout of a flow-soldering plant would be ideal for high-volume production of mobile phones but unsuitable for use in the manufacture of control panels for use in a power station.

Available personnel

The availability of appropriately trained and experienced personnel is also an important consideration. The skills and abilities of personnel need to be taken into account, particularly where new manufacturing technologies and processes are introduced. Where new plant and equipment is to be introduced appropriate training will be needed for existing staff. Alternatively, new staff will need to be recruited and they may also need training. All of this may incur appreciable cost.

Production control

Any production method will require an effective means of control. This should encompass a number of factors, including ensuring the supply of materials and components as well as setting the scale of production and the timing of the different processes required. Effective production control can help to reduce wastage, minimize energy costs and avoid hold-ups and delays in production. The method of production control needs to be considered at an early stage.

Quality control

Quality control helps to ensure that a manufactured product is fit for purpose and that it fully meets the needs and expectations of customers. We will examine this in further detail in Chapter 6 but, for now, you just need to remember that any method of production needs to incorporate a means of ensuring that the quality of production is maintained at a satisfactory level.

Production and manufacturing cost

We have already mentioned some of the costs associated with production. Some of these costs (such as initial capital investment in new plant and equipment) are obvious but others (such as retraining of personnel) are not. When making decisions about what methods of production to use it is essential to have a detailed understanding of the costs involved and to be able to make realistic estimates of what the costs and benefits might be. We will be looking at cost factors next.

Cost factors

> **Key point**
>
> Direct costs are costs that can be directly attributed to the manufacture of a specific product or the delivery of a specific service. Examples are the cost of material used to manufacture the product (such as sheet metal), the cost of energy required to manufacture the product (such as gas or electricity) and the cost of labour required to manufacture the product.

Depending upon the type or production used there will be a variety of different cost factors that need to be taken into account. These include the cost of labour and materials as well as the cost of energy and the overhead costs associated with plant, premises and equipment. To help understand what's going on, cost factors are usually divided into direct and indirect costs. It is import to understand how these two categories of costs differ and why it is essential to be able to distinguish between them.

Direct costs

Direct costs include all costs attributable to a particular product or service. They include the cost of the materials used as well as the cost of labour and energy used to manufacture the product. They do not include overhead costs associated with the premises (e.g. cost of heating and lighting) nor do they include the costs of back office functions such as finance, accounting and personnel.

Indirect costs

Indirect costs are those costs that cannot be directly attributed to a particular product or service. They include back office functions such as finance, accounting and personnel. They also include the cost of company management, research, design and development.

Case study: Pipeline Services Ltd

Pipeline Services Ltd is a company that specializes in the manufacture and operation of remotely operated vehicles (ROV). These provide a means of inspecting and carrying out work in situations where human operators may be exposed to significant risk. Typical situations where an ROV is used include inspection in pipelines, tunnels, and marine oil and gas production. An ROV is a small electrically powered vehicle, often with a cable connection to a control unit fitted with a joystick and other controls that operate grabs and cutting devices. To assist the operator, several cameras are fitted to the ROV with video relayed to display screens on the operator's console.

Key point

Indirect costs are costs that cannot be directly attributed to the manufacture of a specific product or the delivery of a specific service. Examples are the cost of management salaries, factory overheads (such as the cost of heating and lighting), and the cost of sales and marketing.

Figure 5.5 An ROV used in the offshore oil and gas industry.

Figure 5.6 The remote control console for a large ROV.

In order to cater for different types of pipeline, including those where specific hazards might be present, several different types and sizes of ROV are manufactured to meet individual customer requirements. Pipeline Services Ltd manufactures its ROVs by means of batch production methods in quantities of between four and ten units with an identical specification. The company also maintains a stock of replacement parts for each type of ROV and offers on-site maintenance where required. Pipeline Services Ltd has a small team of engineers and designers that meet with new clients, agreeing specifications and drafting contracts. Production starts when contracts have been finalized. Following manufacture and commissioning trials, ROVs usually enter service within six to eight weeks from the start of a project.

Test your knowledge 5.3

Explain why Pipeline Services Ltd use batch production methods for the manufacture of its remotely operated vehicles (ROV).

Test your knowledge 5.4

Classify each of the following costs incurred by Pipeline Services Ltd as either direct or indirect:

1. The cost of metal used to manufacture the chassis of an ROV

2. The cost of labour required to manufacture cable harnesses for use on the ROV

3. The cost of marketing the ROV to the oil and chemical production industries

4. The cost of underwater cameras used on the ROV

5. The cost of building and maintaining an ROV spare parts department

6. The cost of managing an ROV project and agreeing specifications with a new client.

Just-in-time production

Just-in-time (JIT) production is a manufacturing system where the parts and components required for the manufacture and assembly of a product are only replenished at the point at which they are actually needed. This avoids having to hold large quantities of stock parts but it only works when production is closely monitored. JIT manufacturing was pioneered in the Japanese motor industry (Toyota) but has become commonplace where continuous flow production is used. JIT reduces the cost of stockholding but it relies on being able to accurately predict the need for parts and components used in the manufacturing process.

JIT is often preferred for parts and components that might have a higher purchase price or holding cost. Furthermore, over time and due to advances in manufacturing technology, the cost of parts and components may fall significantly.

JIT purchasing is the purchase of parts and materials so that they are only delivered when they are needed for production. Benefits include less money tied up in stockholding and reduced space for storage. JIT also helps to avoid the risk of inventory items becoming outdated or obsolete. On the other hand, buying parts, components and materials in relatively small quantities can be, on a per-unit basis, significantly more expensive than purchasing in bulk (which can often attract significant discounts). For example, the cost of purchasing a small quantity of ten power transistors might be £12 (equivalent to a per-unit price of £1.20) whilst a bulk purchase of 1,000 identical devices might be £650 (equivalent to a per-unit price of 65p). Another disadvantage with JIT is the risk that stocks of parts, components and materials might not be available when they are required. This could have the effect of halting production which in turn might result in an inability to meet customer demand.

Key point

With JIT you purchase the minimum quantity of parts, components and materials that will allow you to meet customer demand.

Test your knowledge 5.5

Which of the following is NOT a feature of JIT manufacturing?

1. Purchasing in smaller quantities with frequent deliveries of parts, components and materials.
2. Need for large amounts of storage space to accommodate stocks of parts, components and materials.
3. Stocks of component, parts and materials are unlikely to be time expired.
4. Bulk discounts from suppliers might not be available.

Push and pull production

Push production (or 'make to stock' production) is a type of production where finished items are accumulated and placed in stock for when they are required. The level of production is not governed by the demand at any time.

Pull production (or 'make to order' production) is the opposite of push production. With pull production a product is only manufactured when an order has been received. In other words, this type of production responds to demand at any point in time.

Clearly, both push and pull types of production have their advantages and disadvantages and, in practice, many companies operate a system of manufacturing that is somewhere between the two extremes. In most engineering companies, Supply Chain Management (SCM) seeks to ensure that stockholding of parts and components purchased from external suppliers is not excessive whilst at the same time maintaining the supply of sufficient materials in order to cope with fluctuations in demand. This is yet another reason why it is important for a manufacturing company to know its market and be aware of changes that could affect the demand for its products and services.

Cellular manufacturing

With cellular manufacturing production is based on a number of cells, each of which has the necessary production equipment to carry out a particular task or range of related tasks, such as dashboard fitting, gearbox assembly, etc. The product moves from cell to cell with each cell responsible for completing part of the manufacturing process. The aim of cellular manufacturing is to move as quickly as possible though the chain of cells, with the ability to manufacture a variety of similar products and with as little waste and down-time as possible.

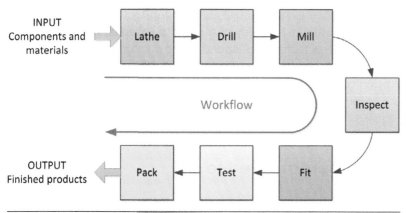

Figure 5.7 Layout of a typical manufacturing cell.

Cellular manufacturing is considered to be a hybrid system that combines the advantages of job shop production with those of a continuous flow assembly line. The cell architecture can easily adapt to change whilst the continuous flow assembly line provides for quick and efficient production. Figure 5.8 shows how these two layouts compare.

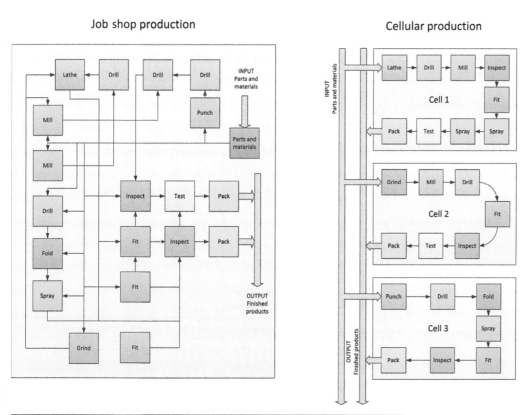

Figure 5.8 Comparison of job shop and cellular production layouts.

Combining the high productivity and throughput of assembly line production with the flexibility of a job shop-orientated cell makes it possible for a factory to respond to the changes necessary for different (but related) products to be manufactured quickly and efficiently. Design modifications and refinements can be carried out at the cell level rather than necessitating change to the entire process.

Test your knowledge 5.6

Explain what is meant by the term 'cellular manufacturing'. State **one** advantage and **one** disadvantage of this method of production.

Learning outcome 5.3

Identify the different stages of production planning

Before attempting to manufacture a new product (or to make substantial changes to a product that is already in production) it is essential to have a plan that relates to each of the different stages of production. Production planning takes into account the scheduling of materials and parts as well as the availability and loading on production equipment and the flow from product initiation to product delivery. An effective production plan offers several advantages including:

- a significant reduction of *work in progress*
- reduction in the inventory of parts and materials held by the company
- more effective utilization of resources (plant, equipment and human)
- accurate prediction of manufacturing time, ensuring that delivery targets are met
- enhanced responsiveness to changes in supply and demand.

Production scheduling

Production scheduling is the process by which manufacturing resources are allocated to the manufacture of a product. Such resources include plant and equipment as well as human resources and the supply of materials and parts. Scheduling is based on an identification of each of the processes involved in manufacture, matching these to the available resources with the ultimate aim of minimizing costs and increasing overall production efficiency. Production scheduling is essentially a time-based view of production and is frequently carried out using dedicated software, planning charts and spreadsheets.

Capacity planning

Once facilities for production have been put in place the next step is to decide how to match the production capacity to meet the predicted demand. In other words, to consider the *loading* on the manufacturing process. Production managers will use a variety of ways to achieve this from maintaining excess capacity to making customers queue or wait for goods to having stocks to deal with excess demand. The process is complex and may require the use of specialized software and forecasting techniques.

Inventory control

With any manufacturing facility good inventory control is an absolute essential. It is estimated that it costs up to 25% of the cost value of stock items per year to maintain an item in stock. Proper inventory control systems have to be used to ensure that there is sufficient stock for production while at the same time ensuring that too much stock is not held. If stock levels are high there are costs associated with damage, breakage, pilferage and storage which can be avoided.

Manufacturing Requirements Planning

Manufacturing Requirements Planning (MRP) is a system used for production planning and inventory control. MRP combines data from one or more production schedules with that from an inventory database and a bill of materials (BOM). MRP makes it possible to forecast the quantities of parts and materials required to manufacture a particular product. An effective MRP system makes it possible to:

* ensure consistent levels of production
* minimize inventory stock levels
* reduce wastage where inventory items might become 'time expired'
* ensure that appropriate quantities of materials are available for production
* maintain optimum production times.

An effective MRP system requires a high level of data integrity. This means that the data held in each of the linked databases is

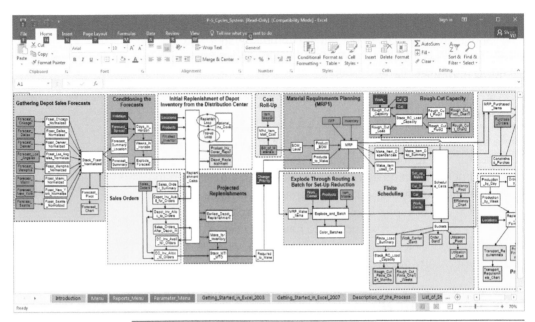

Figure 5.9 A typical MRP spreadsheet.

comprehensive, accurate and up to date (for example, showing current part numbers, supply quantities, stock levels, etc.). Failure to ensure data integrity can lead to an inability to yield reliable information. MRP can be a powerful tool but a great deal of care is required to ensure that it uses data that is current and accurate.

Workforce management and job design

This is related to the need to have a labour force trained to use the facilities installed. The important aspects here are:

- work and method study
- work measurement
- job design
- health and safety.

The production manager has to establish standards of performance for work so that the capacity of a factory can be determined and also so that direct labour costs can be determined. Work study, method study and work measurement activities enable this to be done, as well as helping to promote efficient and safe methods of working. The design of jobs is important in respect of worker health as well as effective work. Good job design can also make the work more interesting and improves employee job satisfaction, which in turn can improve productivity.

Work schedules

A work schedule specifies the resources, processes, plant and materials required to manufacture a product as well as the order and the quantities in which they are required. Work schedules also specify the human resources (personnel) required in order to carry out each of the processes. Work schedules are often created using dedicated software but they can also be based on spreadsheet templates. The following are required in order to produce a comprehensive work schedule:

- detailed engineering drawings and work instructions
- technical data such as specifications, reference tables, tolerance limits, etc.
- suitably trained and experienced personnel
- appropriate plant, equipment, machinery and tools
- components and materials (usually in the form of a bill of materials)
- details of energy requirements, supplies and consumable items (such as fuels and lubricants).

Test your knowledge 5.7

List **three** advantages of having an effective production plan.

Learning outcome 5.4

Explain how to apply typical process charts

Process charts provide you with a simple way of visualizing the manufacturing process. They show the main processes and indicate the flow of products, materials and production. Several different types of process chart are used in engineering, including those that provide an overview of a complete manufacturing process and those that illustrate individual processes within the whole.

To aid understanding, a common set of symbols are used, the most often used being those developed by the American Society of Mechanical Engineers (ASME) shown in Figure 5.10. The five symbols represent the following manufacturing stages:

- *Operation*: a significant process where a product is changed or modified in some way (for example, grinding the surface of a cast light alloy engine block)
- *Inspection*: a check for conformance, quality or quantity (for example, carrying out tests on the taper of a turned steel shaft)

- *Transport*: movement of product, materials or personnel (for example transporting the engine of an aircraft to a hangar where the aircraft will be assembled)
- *Storage*: controlled storage in which parts and materials are stored before use or when a part-finished product is retained for some purpose before being returned to manufacture (for example, where pre-assembled printed circuit boards are stored prior to assembly into a consumer electronic product)
- *Delay* (or temporary storage): where a part-finished product is stored or put aside until a process can be finished (for example, waiting for a polymer coating to harden).

A typical example of a process chart is shown in Figure 5.11.

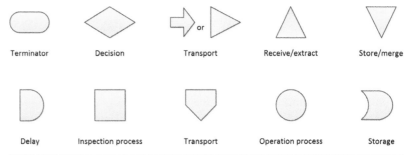

| Terminator | Decision | Transport | Receive/extract | Store/merge |

| Delay | Inspection process | Transport | Operation process | Storage |

Figure 5.10 The American Society of Mechanical Engineers process chart symbols and others.

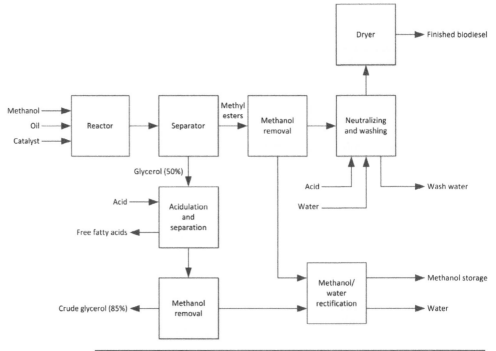

Figure 5.11 An example of a process chart used in the petrochemical industry.

Gantt charts

A Gantt chart is simply a bar chart that shows the relationship of the activities that make up a manufacturing project over a period of time. When constructing a Gantt chart, activities are listed down the page while time runs along the horizontal axis. The standard symbols used to denote the start and end of activities, and the progress towards their completion, are shown in Figure 5.12.

Symbol	Meaning
[	Start of an activity
]	End of an activity
[———]	Actual progress of an activity
[|]	(alternative representation)
V	Time now

Figure 5.12 Gantt chart symbols.

A simple Gantt chart is shown in Figure 5.13. This chart depicts the relationship between four activities A to D that make up a project. The horizontal scale is marked off in intervals of one day, with the whole project completed by day 14. At the start of the sixth day (see *time now*) the following situation is evident:

- Activity A has been completed
- Activity B has been partly completed and is on schedule
- Activity C has not yet started and is behind schedule
- Activity D is yet to start.

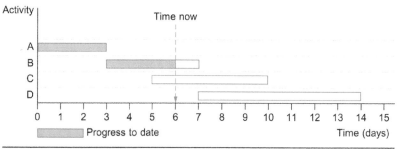

Figure 5.13 A simple Gantt chart.

Another example is shown in Figure 5.14. This chart depicts the relationship between six activities A to F that make up a project. The horizontal scale is marked off in intervals of one day, with the whole project completed by day 18. At the start of the eighth day (again marked *time now*) the following situation is evident:

- Activity A has been completed
- Activity B has been partly completed but is running behind schedule by two days
- Activity C has been partly completed and is running ahead of schedule by one day
- Activity D is yet to start
- Activity E has started and is on schedule
- Activity F is yet to start.

As an example of how a Gantt chart is used in a practical situation take a look at Figure 5.15. This shows how a loudspeaker manufacturer might use this technique to track the progress of a project to produce a new loudspeaker design. The chart shows the situation at the beginning of Day 4 with all activities running to schedule.

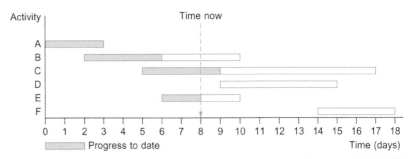

Figure 5.14 Another simple Gantt chart.

Activity	Days									
	1	2	3	4	5	6	7	8	9	10
Draft spec. and budget										
Specify materials/parts										
Design cross-over										
Construct cross-over										
Design enclosure										
Build enclosure										
Order/await drivers										
Assemble components										
Test										

Figure 5.15 Gantt chart for producing a prototype of a new loudspeaker design.

Test your knowledge 5.8

Figure 5.16 shows the Gantt chart for the design and manufacture of a loudspeaker. Assuming that the project is at the start of day 6, which activities are:

(a) On schedule.

(b) Behind schedule by one day, or less.

(c) Behind schedule by more than one day.

(d) Ahead of schedule.

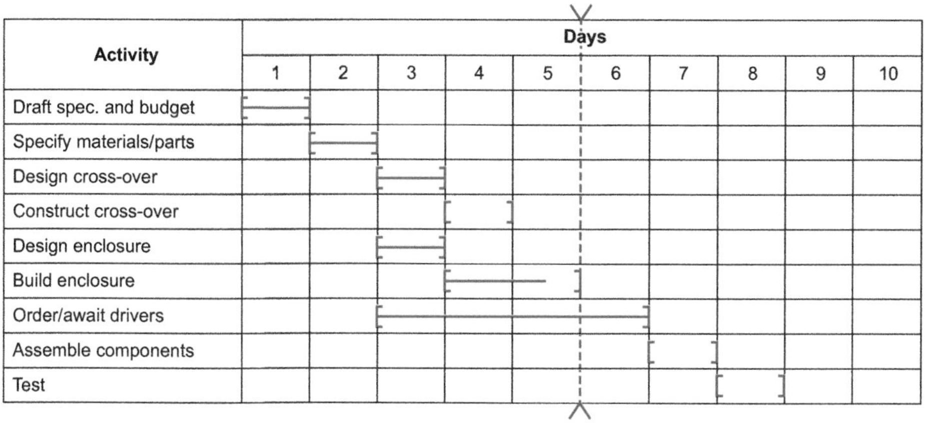

Activity	Days									
	1	2	3	4	5	6	7	8	9	10
Draft spec. and budget										
Specify materials/parts										
Design cross-over										
Construct cross-over										
Design enclosure										
Build enclosure										
Order/await drivers										
Assemble components										
Test										

Figure 5.16 Gantt chart for the design and manufacture of a loudspeaker.

Review questions

1. Describe **three** different methods of production.

2. What method of production would be appropriate for:

   a) a Formula 1 race car?

   b) an aircraft engine?

   c) a small family car?

   d) a domestic microwave oven?

   Explain your answers.

3. List **four** considerations that need to be made when selecting a production method.

4. What is meant by 'job shop production'? Give an example of where this type of production might be appropriate.

5. Describe **two** significant benefits of effective production control.

6. Explain the difference between direct and indirect costs.

7. Give **three** examples of direct and indirect costs associated with the manufacture of a domestic central heating boiler.

8. Explain the differences between push and pull production.

9. Describe **three** advantages of just-in-time (JIT) manufacturing.

10. Explain, with the aid of a diagram, how a Gantt chart is used.

Chapter checklist

Learning outcome	Page number
5.1 Explain the different types and methods of production	114
5.2 Recognize the considerations that need to be made when selecting a production type or method	117
5.3 Identify the different stages of production planning	128
5.4 Explain how to apply typical process charts	131

CHAPTER 6

Quality control and quality assurance

When you have completed this chapter you should understand production activities, including being able to:

6.1 Explain the meaning of the terms 'quality control' and 'quality assurance'.

6.2 Explain the application and content of the BS EN ISO 9000 series of standards.

6.3 Describe the role and stages of inspection activities.

6.4 Explain the role and responsibilities of the Quality Manager.

6.5 List the elements of quality planning.

6.6 Describe the principles of Total Quality Management (TQM).

6

Chapter summary

Being able to deliver a quality product or service is a key objective for the vast majority of engineering businesses. Quality is essentially about knowing what your customer or client wants and expects and then being able to deliver it. This chapter will introduce you to the principles of quality control and the importance of quality assurance. It will also help you to understand the importance of quality standards and how they are implemented in a typical engineering organization.

Learning outcome 6.1

Explain the meaning of the terms 'quality control' and 'quality assurance'

Quality is primarily about being able to meet a customer's expectations. If you can meet a customer's expectations your product or service can be said to be 'of acceptable quality'. If you fail to meet your customer's expectations your product or service will be of 'unacceptable quality'. Of course, different customers might have different expectations so it is vitally important that your product is specified accurately and comprehensively and that your customers know exactly what it is that they are buying.

Problems often arise when a customer is not fully aware of what they are actually purchasing. For example, most customers

Figure 6.1 Vehicle manufacturers are increasingly aware of the need to build quality into their cars.

would expect a service contract to have a specified response time. If this was stated as 'within 24 hours' the quality of service would effectively be the same if the service was delivered 3, 13 or even 23 hours after receiving notice from the customer. That said, the customer is likely to be much happier with a faster response but the longer time could still be within the specified interval forming part of the contract between the service provider and the client.

What is 'quality'?

A Rolls-Royce may be perceived as a 'quality motor car'. Equally a Ford or Volkswagen can be considered 'quality products' *as long as they meet their customers' requirements*. In fact, if you just want a small car for shopping and commuting and you don't have a large budget, a Ford or Volkswagen would actually be a much better choice of car! So what do we mean by 'quality'? Quality is generally defined as 'fitness for purpose' and this means meeting the identified needs of customers. Since it is the customer who judges value received and registers satisfaction or dissatisfaction it is really the customer that determines whether or not a company has produced a quality product.

This does bring problems for manufacturers since customer perceptions of quality vary, some customers will like a product more than other customers will. Hence a manufacturer has to use some more objective criteria for assessing fitness for purpose. This might include:

* design quality
* conformance quality
* reliability
* after-care services.

The first in the list, *design quality*, is the primary responsibility of R&D and Marketing. It relates to the development of a specification for the product that meets identified needs. *Conformance quality* means producing a product that conforms to the design specification. A product that conforms is a quality product, even if the design itself is for a cheap product. That seems contradictory, but consider the following example. A design is drawn up for a 'budget' camera, which is made from inexpensive materials and has limited capability. This camera serves a particular market. If the manufacture conforms to the specification then the product is of high quality, even though the design is of 'low' quality compared with other more up-market cameras.

Quality and reliability

Quality and reliability are related but they are not the same thing. Reliability includes things like continuity of use measured by things like 'mean time between failures' (MTBF). Thus a product will operate for a specified time, on average, before it fails. It should also be maintainable when it does fail, either because it can easily and cheaply be replaced or because repair is fast and easy. After-care relates to after-sales service, guarantees and warranties. We will be looking at this again later in this chapter.

Quality-related activities

Within a manufacturing business, quality-related activities usually include:

- inspection, testing and checking of incoming materials and components
- inspection, testing and checking of the company's own products during and after manufacture
- administering supplier quality assurance systems
- dealing with complaints and warranty failures
- building quality into the manufacturing process.

Some of these activities are performed, after the event, to monitor quality. Other activities may be carried out to prevent problems before they occur. Some activities may also be carried out to determine causes of failure that relate to design rather than manufacturing faults. Once again, we shall be looking at this in further detail later in this chapter.

Quality control

Quality control (QC) should be built into all stages of the product life cycle, from the early stages of design right through to the final delivery to the customer, as shown in Figure 6.2.

Quality assurance

The terms quality assurance (QA) and quality control (QC) are often used interchangeably but they have different meanings. The essential difference between QA and QC is that the former is *process* oriented whilst the latter is *product* oriented. In other words, QA is about the way things are done and QC is about checking that they've been done correctly.

Figure 6.2 Product life cycle with quality control activities.

Table 6.1 Examples of quality control activities at different stages of the product life cycle.

Life cycle stage	Examples of quality control activities
Design	Agreeing specifications, finishes and tolerances with clients (e.g. specifying the thickness of surface coating for a bearing)
Purchasing	Checking that suppliers are able to deliver parts, components and materials of appropriate quality (e.g. by routine sampling of batches of electronic components)
Production planning	Ensuring that quality checks are introduced at key stages during production (e.g. by identifying points at which a product moves from one process to the next and is available for inspection)
Manufacture	During manufacture as processes are applied (e.g. by checking that copper conductors are clean and free from oxidation before attempting to solder joints)
Final inspection	Carrying out a detailed product inspection when all manufacturing processes have been completed (e.g. by inspecting and certifying that the product conforms to a recognized standard or published specification)
Dispatch	Packaging and transportation (e.g. by ensuring that appropriate materials are used and that the package or container is appropriately labelled)
After-care	Providing details of after-sales service, helplines and documentation relating to operation, maintenance and service (e.g. by ensuring that contact details are kept up to date and that online documentation is kept up to date)

Test your knowledge 6.1

List **four** different quality-related activities in a typical manufacturing business.

Test your knowledge 6.2

Explain what is meant by a) design quality and b) conformance quality.

Learning outcome 6.2

Explain the application and content of the BS EN ISO 9000 series of standards

Being able to conform with recognized standards is important for many products. BS EN ISO 9001 is an internationally recognized quality assurance standard that brings together all the activities that may already exist in a company. A *Quality Management System* (QMS) based on ISO 9001 helps an organization to monitor and manage quality on an ongoing basis across all aspects of its operation. It helps companies to benchmark their practices and work towards consistent standards of performance and service.

ISO 9001 offers the following benefits:

- ensures that consistent manufacturing standards are achieved
- improves responsiveness by helping to ensure that customers' needs are met
- improves efficiency and productivity
- motivates and engages staff at all levels within an organization
- indicates a commitment to quality internally and externally
- broadens business opportunities by demonstrating compliance
- provides a 'competitive edge' when marketing and negotiating for new business.

The BS EN ISO 9001 standard sets out how to establish, document and maintain an effective quality system. In order to implement the ISO 9001 standard a company will need to:

- identify the requirements of ISO 9001 and how they apply to the business
- establish quality objectives for the business
- produce a documented quality policy and procedures
- communicate the quality policy and related procedures throughout the business

- develop and implement an effective system of auditing to ensure that systems and procedures are being adhered to
- perform management review meetings to review the business performance and set new targets and action points.

Key point

BS EN ISO 9001 is not a specification. It does not set levels of performance nor does it define the absolute quality of a product or service. ISO 9001 can be applied in a wide variety of different types and sizes of organization, not just engineering manufacture.

Management Responsibility
- Strategic Planning
- Quality Policy
- Management Review

Resource Management
- Production Planning
- Training and Development
- Product Design

Measurement and Improvement
- Quality Audit
- Customer Satisfaction Surveys
- Continuous Improvement

Product Realization
- Inspection
- Quality Control
- Process Control

Figure 6.3 General view of a Quality Management System.

Test your knowledge 6.3

List **four** benefits of a Quality Management System (QMS).

Documentation and procedures

A QMS is usually defined and supported by several documents including a quality manual, process/procedures manual and periodic internal/external audit reports in compliance with ISO 9001.

Quality manual

The quality manual defines the scope of the quality system and describes the processes that underpin the QMS. It usually begins with a clear policy statement together with an explanation of the objectives of the policy. The quality manual will normally provide an overview of the individually documented procedures and the relationship between them.

Process/procedures manual

The process manual will define the responsibility for a process, and how it interacts with other processes within the organization. The process manual will indicate the customer and client requirements as well as the need to demonstrate compliance with industry standards and any applicable legislation. Procedures should relate to:

* control of documents and records
* control of records
* internal audit procedures
* non-conforming products
* corrective action
* preventive action.

Work instructions

Work instructions should be provided that accurately describe how specific tasks are performed. As well as dealing with standard processes, they should take into account any client-specific requirements. Work instructions should be formulated with the assistance of all those involved with actually performing the work and disseminated widely within relevant departments.

Audit reports

Audit reports are essential to a review of the effectiveness of a QMS. An Audit Schedule needs to be drawn up by senior management (with input from the Quality Manager and Quality Team). Reports need to be circulated to those responsible for carrying out a process and constructive feedback should be welcomed from staff at all levels. When responding to a report, *action points* should be agreed and those responsible should be named together with an indicator of the timescale. Senior management should be kept fully informed so that quality improvement is monitored at the highest level within the organization.

Test your knowledge 6.4

In relation to the operation of a Quality Management System (QMS), explain the purpose of a) a quality manual and b) an audit report.

Learning outcome 6.3

Describe the role and stages of inspection activities

We've already mentioned that inspection plays an important part in quality control and that it needs to be built into the production process at various stages. The purpose of inspection is to check that:

- any deviation from what is expected is detected
- adjustments can be made to ensure the final quality is in line with customers' expectations.

Inspection is carried out at various stages including:

- goods inward (as components and materials are received)
- during production (often as part of an activity referred to as *process control*)
- final inspection (prior to dispatch to the customer or prior to delivery).

Key point

Inspection aims to determine if there has been any deviation from what is expected and is the main tool by which quality control is applied.

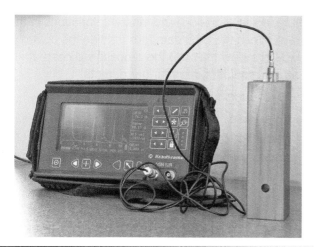

Figure 6.4 Ultrasonic testing equipment used during the inspection of aircraft parts.

The role of the Quality Inspector

The Quality Inspector (or any other person responsible for carrying out an inspection) checks compliance with the company's established quality standard and procedures. The inspector usually follows a pre-established checklist based on the detailed product specifications. As mentioned previously, inspection can be carried out at various stages including the materials, components and parts used for production and on part-finished products as well as the final product prior to shipment to the customer.

Statistical process control

Statistical process control (SPC) is as a means of checking and ensuring that quality/performance is within required limits. SPC uses statistical methods to monitor and control specific aspects of a product during production. It aims to improve consistency and reduce waste and it can be applied to any process where the conformance of a product can be measured (for example, physical dimensions, electrical conductivity, thermal conductivity, etc.).

SPC relies on data obtained during the manufacturing process. These measurements are made in real time during the production process. The measurements are recorded and usually plotted on a graph (a *control chart*) with pre-defined control limits. In most cases there is an *upper control limit* (UCL) and a *lower control limit* (LCL). These should not normally be exceeded and when they are they indicate a defect condition in which a product should be rejected for scrap or reworked if it is possible and economically viable to do so.

When the process is operating as expected, the SPC data will fall between the UCL and LCL. Some variation will occur within the limits due to variables within the process (such as those induced by temperature, noise, vibration, and inherent variations in the materials used). When the SPC data falls outside the control limits (above the UCL or below the LCL) the process will need attention in order to avoid the manufacture of further defects.

The benefits of introducing an SPC system include:

- improved yield and a significant reduction in scrap
- improved cost-effectiveness of the production process
- ability to detect variations as and when they occur
- any unwanted process changes are detected immediately.

Test your knowledge 6.5

Explain what is meant by statistical process control (SPC).

Test your knowledge 6.6

List **three** benefits of statistical process control (SPC).

Case study: Warren RF Technology

Warren RF Technology manufacture a variety of components used in the telecommunications sector. These include a range of terminating load resistors used for carrying out measurements on low-power radio systems. Most popular in this range is a 50 ohm resistor encapsulated within a low-loss coaxial SMA connector (see Figure 6.5).

To improve consistency of their resistor products, Warren RF Technology use a laser device to accurately trim the resistance of the load so that its value is close to the nominal 50 ohm. To assist and inform the production process, the company has invested in SPC and the data from the output of the laser trimmer is used as one of the inputs to the system.

Warren RF technology's customers require a component that is accurate to within ±5% of the nominal 50 ohm value but the company has decided to improve on this specification by producing a component that is accurate to ±4%. This means that the resistance of a finished component should lie within the range 48 to 52 ohm with a mean value of 50 ohm.

A control chart for the laser trimming process is shown in Figure 6.6. Note that the upper and lower specification limits (USL and LSL respectively) are entered into the spreadsheet and the UCL and LCL are marked in red on the chart. The mean value of resistance is marked in blue.

Figure 6.5 A 50 ohm terminating resistor manufactured by Warren RF Technology.

Control Chart for Mean and Range

Laser trimmer
05/06/2016

Quality Characteristic	Average Resistance (Ohm), X-bar
Sample Size, n	5
k	3

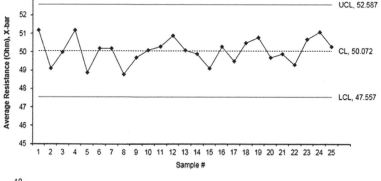

Statistics from Data Table

R-bar	4.360
Process Mean, μ-hat	50.072
Process St.Dev., σ-hat	1.874
$\sigma_{X\text{-}bar}$	0.838

Process Capability

Upper Spec Limit, USL	52
Lower Spec Limit, LSL	48
C_p	0.356
CPU	0.343
CPL	0.368
C_{pk}	0.343
Percent Yield	71.37%

Control Limits for R Chart

CL_R	4.360
UCL_R	9.219
LCL_R	0.000

Control Limits for X-bar Chart

$CL_{X\text{-}bar}$	50.072
$UCL_{X\text{-}bar}$	52.587 $CL+k\sigma_{X\text{-}bar}$
$LCL_{X\text{-}bar}$	47.557 $CL-k\sigma_{X\text{-}bar}$
α	0.0027
ARL	370.4 samples

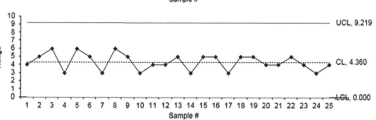

Figure 6.6 An SPC control chart for the laser trimming process.

Activity 6.1

Use the control chart shown in Figure 6.6 to answer the following questions:

1. What is the average value of resistance of the load resistors produced by the laser trimming process?
2. What is the average deviation of resistance from the nominal value?
3. What are the upper and lower control limits (UCL and LCL) for the laser trimming process?
4. How do the values of UCL and LCL compare with the upper and lower specification limits (USL and LSL respectively)?
5. Of the 25 resistance values recorded in the SPC chart, how many are above and how many are below the nominal resistance value?

Document control

Document control is an important part of any quality system. Various documents are needed for quality assurance and the control of those documents is vitally important to ensure that they provide a written framework within which the QMS operates. They also provide evidence that the system exists and that it is fit for purpose. In fact, it is not possible to be confident that the system is operating correctly without referring to its documented procedures.

There are several main types of documents that need to be controlled. In most organizations these include:

- *Quality procedures*: A detailed written description of the quality system and the controls that are in place within the company (these often form part of a quality manual or are documented separately as a procedures manual).
- *Work instructions*: A description of a particular operation or task in terms of what must be done, who should do it, when it should be done, and what materials and processes should be used. In many cases, a series of work instructions are used to describe each stage in the production or manufacturing process.
- *Test specifications*: A detailed list of characteristics and features used to verify conformance with the design specification together with details of any measurements that are to be made out and how they should be carried out.

Traceability

Documents are often used to assist in the process of identification and traceability of products, components and materials. This is vitally important in critical sectors such as aerospace, nuclear and chemical engineering. Traceability is essential when it becomes necessary to eliminate the causes of non-conformance. Traceability is achieved by coding items and maintaining records that can be updated throughout the working life of a component part.

Test your knowledge 6.7

Explain why traceability is important for certain engineering components and products. Give **two** engineering sectors where traceability is important.

Key point

Document control involves ensuring that documents are comprehensive, relevant and up to date. Document control can provide evidence that a quality system is in place and that it is effective.

Quality procedures

Effective quality procedures are essential in any manufacturing context. Without them, excessive variations in performance can go undetected. Quality procedures need to underpin and inform quality controls. For example, a sample of the product might need to undergo certain tests not only to ensure that the product meets the stated specification (and is therefore 'fit for purpose') but also to confirm that the manufacturing process is operating as expected. Depending on the product and the nature of the processes that are applied to it during manufacture, quality procedures might involve a number of different stages and processes, including different methods of sampling and quarantine procedures.

Quarantine

In many production processes defective work is removed and put aside for detailed inspection. In order to ensure that rejected items do not re-enter the manufacturing process (or are not shipped to customers) defective parts are usually relocated to a clearly marked quarantine area that is physically separate from the main production process. A decision can then be made to scrap, rework or adjust the work in some way. A detailed examination can often reveal the root cause of a defect (such as a faulty machine) and this, in turn, can be useful in informing changes to the production process.

Sampling

Sampling is a procedure that is frequently used to check production processes. Sampling can be carried out in various ways:

- *Spot checks:* Samples of production are withdrawn for inspection as and when required. Spot checks are usually carried out on an infrequent basis and they are often used when there might be specific concerns about the production process (for example, when a machine tool is replaced or when a new batch of materials is brought into use).

- *Random sampling:* Random sampling occurs on an irregular and random basis. Random sampling often involves taking items from the production output at random intervals, checking for consistency with the results from previous samples.
- *Process sampling:* Process sampling involves taking a sample of the output from a particular process and comparing it with a sample taken from a different process. Process sampling is particularly useful where processes are applied in parallel, for example, when alternative protective coatings can be applied to a product).
- *Batch sampling:* In batch sampling a limited number of items are taken from a production batch and checked for consistency with those taken from other batches. Batch sampling can reveal time-related problems within the production process. For example, progressive degradation of cutting and grinding tools.

Failure

Failure is an inability of a product to carry out its specified function. Failure of a product overall will often result from the failure of an individual component. Since engineered products often use a very large number of components, the reliability of the product will be significantly reduced as the number of component parts increases.

A product can fail in many ways, including misuse and inherent weakness in its design. It can also fail suddenly or gradually. Sudden failures are those that cannot be anticipated (even after a close examination of the product prior to failure). Gradual failures can normally be detected when they occur, usually by a loss of performance or an inability of the product to meet some aspect of its technical specification. Failures may also be partial or total, in which case they are referred to as catastrophic.

Production engineers are often concerned with how many manufactured products or component parts are faulty or become early failures due to some manufacturing defect. Such defects can usually be rectified by making modifications or improvements to the production process or by ensuring that the materials and components used are of better quality.

We've already seen that being able to detect trends involves gathering statistical data on the number of products manufactured and the number of defective products or early failures. The number of defects detected often varies with time because different batches of materials and components are used. For this reason, hourly, daily, weekly or monthly production data is gathered and then examined in order to detect the proportion of failures. Once we have this information we can start to examine the causes and make modifications and improvements that will rectify the problem.

Table 6.2 shows the production of smoke detectors manufactured by a particular company in the second quarter of 2016.

Table 6.2 Production data for smoke detectors.

Second quarter 2016	April	May	June	Second quarter totals
Total production	29,700	45,150	51,000	125,850
Defective	495	525	451	1,471
Percentage faulty	1.66%	1.16%	0.88%	1.17%

The trend indicated by these statistics is that, whilst the production volume has increased the proportion of faulty smoke detectors has fallen by about 50% in the three-month period. This indicates a significant improvement in quality.

Test your knowledge 6.8

Explain the purpose of a quarantine area in a production environment.

Reliability

Reliability is important in all engineered products and since many products rely on the correct functioning of a large number of component parts reliability will be impaired unless the reliability of each component part can be guaranteed. Suppose, for example, that it is known that one component out of 1,000 would break down every hour. A product using 50 of these components would break down at an average interval of 20 hours.

To get some idea of the importance of reliability as a topic for engineers, consider each of the following:

- Modern aircraft use complex fly-by-wire systems that rely on the correct operation of computer systems which use thousands of individual components. Without these systems the aircraft could not remain in the air.
- In the UK the Air Registration Board will only license aircraft to use an instrument landing system if the fault rate for the total system is less than one in ten million.
- In the case of military aircraft on a combat mission, or a missile flight, the consequences of a failure could potentially jeopardize the entire mission. It has been estimated that unreliability

Key point

Key point

The reliability of a product or component is expressed in terms of the probability that it will perform its required function under stated conditions for a stated period of time.

Key point

The rate at which a product or component fails (i.e. its failure rate) is an indicator of its reliability.

costs the RAF more than £100 million each year in spares and maintenance costs.

- If the system used for controlling a chemical or nuclear plant should fail, there could be very serious consequences including the possibility of an environmental disaster that could affect millions of people and cause pollution that would last for hundreds of years.
- If a large-scale production process fails, the cost of shutdown may be considerable in terms of spoiled production and loss of output. The loss of production of a major plant can often amount to more than £10,000 per hour.

Whilst reliability is an important factor in the safety of some products, it is by no means the only factor. A system or equipment can do its job and still be unsafe. Some major disasters have been caused by human failure, not by mechanical or electronic breakdown. In an engineering context, reliability is usually defined as the ability of a product or component expressed in terms of the probability that it will perform its required function under stated conditions for a stated period of time.

Figure 6.7 Reliability is important where a product has many thousands of interdependent parts.

The bath tub diagram

When an engineered product first enters service early failures may occur. These may be caused by manufacturing faults, design faults or misuse. The early failure rate may, therefore, be relatively high, but falls as the weak parts are replaced. There is then a period during which the failure rate is lower and fairly constant, and finally the failure rate rises again as parts start to wear out. We often refer to this as a bath tub diagram because of the shape. Although the steady rate (at the bottom of the bath tub diagram shown in Figure 6.8) is often shown as a straight line, in practice it will be wavy and in good (reliable) products it may be a very long time before the wear-out period is reached. The part of main interest is the constant failure rate period. The three distinct regions of the diagram are defined as follows:

- *early failure period:* the early period during which the failure rate is decreasing rapidly (we sometimes refer to this as the burn-in period because the high initial failure rate may result from components that were defective in manufacture)
- *constant failure rate period:* the period during which failure occurs at an approximately uniform rate
- *wear-out failure period:* the period during which the failure rate of some items is rapidly increasing as the product reaches the end of its working life due to progressive deterioration.

Note that the early life period for many products often extends well beyond any factory burn-in and since such failures will occur during the early stages of customer ownership they will often be viewed far more negatively than equivalent failures in later life. For example, we might tolerate the windscreen wiper failure on our car after a period of a year but most of us would be far less tolerant if the failure occurred while we were demonstrating the new car to our friends during the first week we owned it. However, many products are only used occasionally and in this case the early life period could be well in excess of a year and in some situations could last longer than the full working life of the equipment. A windscreen wiper may operate for typically no more than 1,000 hours during the full life of a car which could be more than ten times greater.

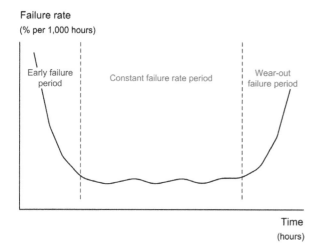

Figure 6.8 The bath tub diagram.

An important measure of the reliability of a product is how long it can be expected to operate before a failure occurs. Depending on the type of product, and whether it is repairable or not, this is defined on one of two ways:

1. *Mean time between failures (MTBF):* this applies to *repairable* items, and means that if an item fails, say, five times over a period of use totalling 1,000 hours, the mean (or average) time between failures would be 1,000 divided by five or 200 hours.
2. *Mean time to failure (MTTF):* this applies to *non-repairable* items, and it means the average time an item may be expected to function before failure. It is found by stressing a large number of the items in a specified way (e.g. by applying certain electrical, mechanical, heat or humidity conditions), and after a certain period dividing the length of the period by the number of failures during the period.

MTBF and failure rate

When the failure rate is constant (as it is at the bottom of the bath tub diagram during most of the normal working life of a product) the MTBF is equal to the inverse of the failure rate. Take the case of a telephone answering machine with an MTBF of 1 million hours that is used 24 hours per day:

MTBF = 1 / failure rate

Failure rate = 1 / MTBF = 1 / 1,000,000 hours or 0.000,001 failures per hour

This is a tiny figure so, if we use 1,000 hours as a more practical time period, this would give:

Failure rate = 0.001 failures / 1,000 hours or 0.1% per 1000 hours.

Now, since there are 8,760 hours in a year the failure rate of our telephone answering machine (connected and used 24 hours a day) would be:

Failure rate = 0.876% per year.

Note that failure rates are often measured under ideal conditions, usually as a result of tests and measurements made in a laboratory. In the real world they are significantly increased by a number of factors determined by the environment and the way in which a product is used. These factors include temperature, humidity, vibration, etc. Temperature is often the most critical factor that will reduce the reliability of a product. For example, when the surrounding temperature increases by 15°C, the reliability of a hard disk drive is reduced by about 50%.

Service life

Some manufacturers use the term 'service life' to describe how long their products may last in use. This should not be confused with MTBF or MTTF. The reason for this is that the normal life failure rate is not the same as the failure rate when a product begins to wear out. Taking the case of the telephone answering machine mentioned earlier which had a failure rate of 0.876% per year – this would be the same as an MTBF of 114 years, which might imply that the service life is also 114 years. But, is it actually reasonable to expect this product to work for 114 years? Quite apart from the obsolescence of the unit after such a long period it would actually have entered the wear-out period long before this time. Most of the components from which the product is manufactured would have failed long before this!

Key point

The reliability of a product or component is expressed in terms of the probability that it will perform its required function under stated conditions for a stated period of time.

Key point

Burn-in refers to the operation of a product prior to delivery to a customer. Burn-in is used to identify and reject early failures due to the use of defective components or resulting from faulty manufacture.

Figure 6.9 A hard disk drive with a quoted MTBF of 250,000 hours (equivalent to a failure rate of 0.4% per 1,000 hours, or approximately 3.5% per annum).

Key point

Burn-out refers to the end of the working life of a product during which its failure rate increases rapidly as it becomes worn out.

Test your knowledge 6.9

1. A marine radar system has a mean time between failures (MTBF) of 120,000 hours. Determine the failure rate of the radar system expressed as a percentage per 1,000 hours.
2. A low-energy lamp has a constant failure rate of 2.5% per 1,000 hours. Determine the MTTF of the lamp.

Test your knowledge 6.10

The service life (in hours) of a batch of 20 low-energy lamps is as follows: 3,223, 3,001, 5,144, 9,581, 5,376, 5,585, 8,236, 6,959, 2,176, 3,651, 4,889, 4,615, 8,657, 5,900, 6,065, 6,882, 7,125, 7,728, 7,291 and 6,337.

Determine the average service life for this batch and use it to determine the failure rate.

Activity 6.2

Factory tests on a number of identical computer systems reveal the causes of failure shown in Table 6.3. Enter this data into a spreadsheet and present this information in the form of a labelled pie chart. Which component can be considered a) the most reliable and b) the least reliable. If the MTBF of the power supply is quoted as 80,000 hours, what is its failure rate expressed as a percentage per 1,000 hours?

Table 6.3 Causes of failure in identical computer systems.

Component	Number of failures attributable
Hard disk drive	21
Memory module	9
Miscellaneous	3
Motherboard	1
Network card	2
Power supply	5
Processor	11

Figure 6.10 Electrical wiring and connectors fitted to this Rolls-Royce Trent Airbus A380 engine needs to use component parts with very low failure rates.

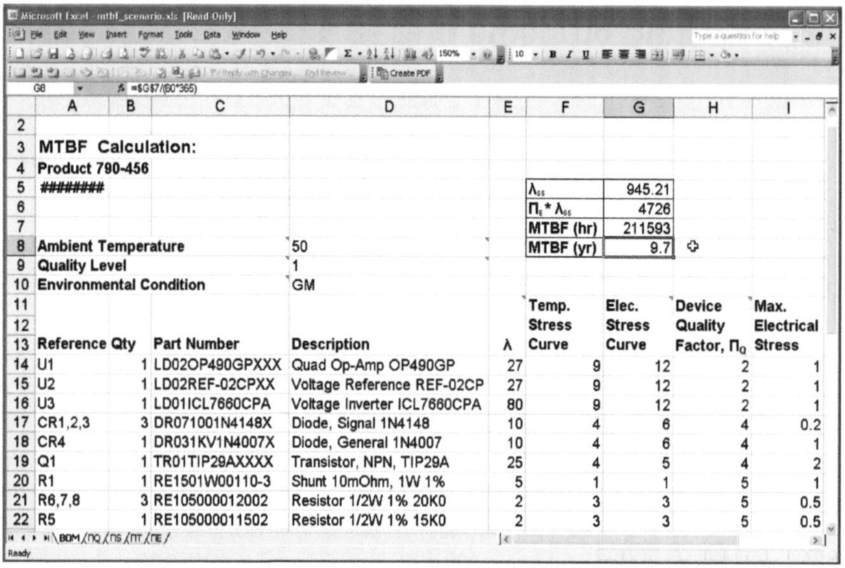

Microsoft Excel - mtbf_scenario.xls [Read-Only]

G8 =G7/(80*365)

MTBF Calculation:

Product 790-456

Ambient Temperature	50
Quality Level	1
Environmental Condition	GM

λ_{SS}	945.21
$\Pi_E * \lambda_{SS}$	4726
MTBF (hr)	211593
MTBF (yr)	9.7

Reference	Qty	Part Number	Description	λ	Temp. Stress Curve	Elec. Stress Curve	Device Quality Factor, Π_Q	Max. Electrical Stress
U1	1	LD02OP490GPXXX	Quad Op-Amp OP490GP	27	9	12	2	1
U2	1	LD02REF-02CPXX	Voltage Reference REF-02CP	27	9	12	2	1
U3	1	LD01ICL7660CPA	Voltage Inverter ICL7660CPA	80	9	12	2	1
CR1,2,3	3	DR071001N4148X	Diode, Signal 1N4148	10	4	6	4	0.2
CR4	1	DR031KV1N4007X	Diode, General 1N4007	10	4	6	4	1
Q1	1	TR01TIP29AXXXX	Transistor, NPN, TIP29A	25	4	5	4	2
R1	1	RE1501W00110-3	Shunt 10mOhm, 1W 1%	5	1	1	5	1
R6,7,8	3	RE105000012002	Resistor 1/2W 1% 20K0	2	3	3	5	0.5
R5	1	RE105000011502	Resistor 1/2W 1% 15K0	2	3	3	5	0.5

Figure 6.11 Computer software is often used to predict equipment failure rates by taking into consideration the failure rate of each individual component and applying weighting factors for various environmental and operational stresses.

Activity 6.3

ACE Microcontrollers has asked you to advise them on problems that they have identified with the automated production of a computer board. They have provided you with the production data shown in Table 6.4.

Calculate the percentage of boards that fail for each period and for each machine, then compare the statistics and identify any trends. What do they indicate? What can you say about the two machines? What recommendations would you make and what further investigations should the company undertake?

Table 6.4 ACE Microcontrollers production data.

		Week 33	Week 34	Week 35
Machine A	Total production	375	282	467
	Faulty units	22	12	10
Machine B	Total production	230	312	411
	Faulty units	4	10	27

Explain the role and responsibilities of the Quality Manager

The Quality Manager has a key role in many engineering and manufacturing organizations. They will:

- ensure that processes needed for the Quality Management System (QMS) are established, implemented and maintained
- report to senior management on the performance of the QMS and highlight the need for improvement as and when it arises
- promote an awareness of customer requirements and expectations and ensure that they are understood on a company-wide basis
- liaise with the external quality assessment bodies on all matters related to external accreditation process (e.g. ISO 9001)
- ensure that a document control procedure is adopted to approve, review and update all changes to critical documents within the scope of the QMS
- ensure that records are established and maintained to provide evidence that the QMS is being followed and that there is a system in place for the identification, storage, protection, retrieval, retention time and disposition of such records
- ensure that the performance of the QMS is reviewed at planned intervals to ensure its continuing suitability, adequacy and effectiveness
- ensure that quality objectives are set by top management for measuring the performance of the QMS and that these are regularly reviewed
- ensure that all new staff are inducted into the requirements of the QMS related to their own roles and responsibilities, and provide update training as necessary
- ensure that all suppliers used by the organization are selected, evaluated and periodically re-evaluated and that records of this assessment are maintained
- ensure that top management undertake periodic but regular assessments of customer satisfaction and that consequent improvements are identified and implemented
- ensure that an internal audit programme is adopted to verify that the QMS conforms to planned arrangements, QMS arrangements and is effectively implemented and maintained, and ensure that appropriate action is taken when this is not the case
- analyse data on the effectiveness of the QMS and evaluate where continual improvements of the QMS can be made. This

should include data generated as a result of monitoring and measurement and from other relevant sources

- co-ordinate continual improvements of the QMS, ensuring that evidence of corrective and preventive actions taken are recorded and reviewed.

Clearly this can be a very demanding role and it is one that requires well-developed inter-personal skills. A Quality Manager will usually seek to work with the support of a Quality Team with representatives drawn from all areas within the organization. These 'quality representatives' will be charged with embedding and implementing quality procedures at a local level within the organization. They have an important role in ensuring that change is implemented at all levels and in all departments of an organization.

Test your knowledge 6.11

Describe the role of a Quality Manager in an engineering company.

Learning outcome 6.5

List the elements of quality planning

A quality plan is an important document or group of related documents that specify the quality standards, manufacturing processes and detailed specifications relevant to a particular product or service. A typical quality plan will:

- establish quality requirements (i.e. the customer expectations)
- identify individual and collective responsibilities (at all levels)
- specify production times and processes (and the sequence in which they are applied) that will ensure that quality standards will be met
- identify budgets and resources to support quality activities
- establish systems that will measure quality and generate reports
- identify control measures and ensure that inspections and measurements are valid and relevant and results are recorded
- ensure that equipment used for measurement and testing is calibrated periodically
- outline the corrective actions that will be taken if non-conformity is found
- record changes to a product, to its specification, or to the processes used in its manufacture.

From the above you should realize that a quality plan is a complex document (or set of documents) and when initially developed it is essential to be fully aware of the objectives of the plan (i.e. to improve yield, reduce defects, improve reliability, ensure accuracy, and so on). Knowing what the customer expects of the product or service is key in all of this. The quality plan should also reflect the overall mission and strategic objectives of the business.

Deployment and documentation

To be effective, a quality plan needs to be widely disseminated and it needs to be 'owned' by those responsible for carrying it out. Strategic-level quality plans are developed and deployed through a company's planning process. These broad-based quality plans become the guideline for each department's supporting quality plan. This plan should originate at a local (departmental) level but, where appropriate, each function within a department should develop and internally deploy its own operating-level quality plan. Such plans are closely related to work instructions and may even replace them in some cases. The documentation included in such a plan might include detailed drawings, a copy of the customer's order, references to applicable standards, practices, procedures, and work instructions with details of how to produce the specific product or service. These documents can be further augmented by inspection reports, SPC charts, and copies of shipping documents and any certification required by the customer.

Test your knowledge 6.12

Describe **four** different items that should form part of a company's quality plan.

Learning outcome 6.6

Describe the principles of Total Quality Management (TQM)

Total Quality Management (TQM) is a management approach to quality improvement that is embedded in the culture of the organization as a whole. TQM originated more than half a century ago and it provides a means of ensuring that things are done correctly first time, and every time afterwards. TQM has the potential to change attitudes and behaviour, thereby making a positive contribution to a business as a whole. TQM focusses attention on quality in all aspects of the company's operations.

Key principles of TQM

The key principles of TQM are:

1. A management commitment to quality in all areas.
2. Empowering employees by participation, teamwork, training, appraisal and recognition.
3. Decision-making based on reliable and robust methods such as statistical process control (SPC) or Failure Mode and Effects Analysis (FMEA).
4. A commitment to continuous improvement in order to maintain and improve performance standards.
5. A focus on customer needs and supplier partnerships.

Continuous improvement

TQM looks for ways of improving an organization on an ongoing basis. TQM treats an organization as a collection of inter-related processes. It requires that businesses must strive to continuously improve these processes by building on the knowledge and experiences of those involved.

Advantages of TQM

Many large organizations have reported the benefits of introducing aspects of TQM, particularly where managers and employees at all levels work together to address issues that relate to quality. Specifically, this helps to harmonize customer needs and expectations with the organization's own strategic goals and objectives. TQM has been successfully implemented by Ford, NXP Semiconductor, SGL Carbon, Motorola, Toyota and many other companies.

The benefits of TQM include:

- making an organization more competitive
- establishing a new culture that will enable growth and longevity
- providing a working environment in which everyone can succeed
- reducing stress, waste and internal friction
- building effective teams, enabling partnerships and fostering co-operation.

Implementation of TQM

Although originally applied to manufacturing operations, and for a number of years only used in that area, TQM is now used as a

Key point

The term 'total quality' refers to the culture in an organization where quality is paramount and is embedded into the organization as a whole rather than just one or more parts of it.

generic management tool that is equally applicable in service and public sector organizations. TQM aims at achieving continuous improvement to the operation of a business and it permeates through all aspects of work. In order to successfully implement a TQM strategy, an organization needs to have a track record of responsiveness to the environment in which it works. This is an essential pre-requisite since, without it, most organizations fail to embrace the level of change required. In such cases it can be preferable to delay the implementation of TQM until the organization is in a state in which it is likely to succeed.

Test your knowledge 6.13

List **four** advantages of introducing TQM in a manufacturing context.

Review questions

1. Explain the difference between 'quality control' and 'quality assurance'.

2. List **four** quality advantages of introducing an ISO 9001-based Quality Management System.

3. In relation to a Quality Management System, describe the role of a process manual.

4. Explain why sampling might be needed when materials and components first arrive at a manufacturing site.

5. Explain the use of confidence limits in statistical process control.

6. Distinguish between spot checks, random sampling and batch sampling in a manufacturing context.

7. A particular type of heater has a failure rate of 3.9% per 1,000 hours. What is the heater's MTBF and, on average, how long will it be before a heater needs repair?

8. Explain why MTTF is not the same as 'service life'.

9. List **five** advantages of a Quality Management System.

10. Briefly explain the key principles of Total Quality Management.

Chapter checklist

Learning outcome	Page number
6.1 Explain the meaning of the terms 'quality control' and 'quality assurance'	138
6.2 Explain the application and content of the BS EN ISO 9000 series of standards	142
6.3 Describe the role and stages of inspection activities	145
6.4 Explain the role and responsibilities of the Quality Manager	158
6.5 List the elements of quality planning	159
6.6 Describe the principles of Total Quality Management (TQM)	160

CHAPTER 7

Organizational improvement techniques and competitiveness

Learning outcomes

When you have completed this chapter you should understand organizational improvement techniques and competitiveness, including being able to:

7.1 Explain the meaning of the terms lean manufacture, kaizen, just-in-time and kanban and their overall advantages.

7.2 Recognize the need for continuous improvement to ensure organizational competitiveness.

7.3 Recognize the importance of improving productivity.

7.4 Recognize how to manage the production process.

7.5 Recognize the importance of teamwork and the individual's contribution to effective team-work.

Chapter summary

Improving an organization's effectiveness is important for all businesses, large or small. Responding to this challenge is an ongoing commitment and many manufacturing concerns have introduced initiatives that seek to improve effectiveness, enhancing productivity and increasing competitiveness. This chapter looks at a range of solutions including lean manufacturing, just-in-time (JIT) production, and the cycle of continuous improvement activities used by many of today's most successful companies.

Learning outcome 7.1

Explain the meaning of the terms lean manufacture, kaizen, just-in-time and kanban and their overall advantages

Lean manufacturing was first implemented by a Japanese company, Toyota, in an effort to eliminate automotive production line waste. Lean manufacturing can be applied in various ways and assisted by various tools designed to manage and monitor the efficiency of the production process. The ultimate aim of lean manufacturing is that of maintaining consistent manufacturing quality whilst at the same time reducing waste, streamlining workflow and improving the efficient use of production resources. In effect, lean manufacturing focusses on getting the right things to the right place at the right time in the right quantity to achieve optimum workflow, while at the same time minimizing waste and being flexible and adaptable.

What is waste?

We've already said that it's important to eliminate waste. That said, the meaning of the term 'waste' needs to be clarified because, in a production environment, waste occurs in various forms, not just unused material. In broad terms waste can include:

- *over-production* (e.g. manufacturing more items than are currently required by customer demand)
- *unnecessary processing* (e.g. applying a finish that the customer did not request and does not want)
- *time wasted* when production resources (human as well as physical) are idle (e.g. waiting for materials to be delivered or for work to be finished before the next process can start)
- *unnecessary movement* of people or parts during a production process (e.g. part-finished product being transferred to a storage area before later returning back to where it came from)

- *defective products* resulting from poor quality, inconsistent manufacturing (e.g. component parts rejected due to an inability to meet manufacturing tolerances)
- *unused inventory* (e.g. materials and parts that are held in stock and have not had any value added to them)
- *unwanted transportation* (e.g. transporting finished or part-finished goods into an off-site storage depot before eventually dispatching them to customers).

Waste and quality

It is important to associate high quality with low wastage, and vice versa. It is also worth remembering that high quality is linked to high value in the eyes of the customer, so lean manufacturing can bring benefits to the customer as well as to the manufacturer. Lean manufacturing is also based on the principle of giving the customer exactly what he or she wants and nothing more. In other words, all unwanted or redundant features are removed from the product and the manufacturing process just concentrates on delivering what the customer is willing to pay for.

Key point

Lean manufacturing achieves optimum workflow by getting the right things to the right place at the right time and in the right quantity.

Figure 7.1 The ultimate aim of lean manufacturing is that of maintaining consistent manufacturing quality whilst at the same time reducing waste, streamlining workflow and improving the efficient use of production resources.

Test your knowledge 7.1

A manufacturer will only apply a protective coating to an item when specifically requested by a customer. Explain why this is in line with the principles of 'lean manufacturing'.

Kaizen

Kaizen is a commonly used tool that encourages continuous improvement in quality, technology, processes, productivity, safety, and workplace culture. Kaizen focusses on applying small, daily changes that result in major improvements over time. Kaizen is a strategy in which employees at all levels of a company work together proactively to achieve regular improvements to the manufacturing process. Kaizen has been used by many companies to improve competitiveness. As its name might suggest, kaizen was first introduced in Japan and the work 'kaizen' is a combination of two words 'kai' (meaning improvement) and 'zen' (meaning good).

Just-in-time (JIT) manufacturing

Just-in-time (JIT) has already been introduced in Chapter 5. JIT production is an important aspect of lean manufacturing simply because the materials, parts and components required for the manufacture and assembly of a product are only replenished at the point at which they are actually needed. Benefits of JIT include less money tied up in stockholding coupled with a significant reduction in the space required for storage. JIT also helps to avoid the risk of inventory items becoming outdated or obsolete.

Test your knowledge 7.2

Explain **three** key benefits of JIT manufacturing.

Kanban

The Japanese word 'kanban' means 'signboard' or 'billboard'. Kanban boards and charts are visual tools used for managing projects and processes. Kanban is widely used for production and it is particularly useful for JIT manufacturing. Its use is not restricted to manufacturing and it is equally applicable to a wide range of businesses.

Kanban enables teams and organizations to visualize their work, ensuring that resources are deployed in a way that avoids bottlenecks and delays in the production process. It can significantly improve the efficiency of the environment in which it is used, reducing down-time and minimizing waste.

By making production processes (and the links that exist between them) highly visible to all those involved, it is possible to accurately track workflow and improve communications between those involved. This, in turn, ensures the optimum use of resources.

Key principles of kanban

The four key principles of kanban are:

1. *Visualizing work processes*: making work processes visible so that it is possible to identify delays and bottlenecks. This improves communication and ensures that everyone is actively involved.

2. *Limiting work in progress*: streamlining workflow by reducing the amount of work that is not required immediately or is in an idle state. This helps to avoid queues and overloaded resources and is good practice in the context of just-in-time (JIT) manufacturing.

3. *Focussing on workflow*: ensuring that the available resources are deployed efficiently. This smooths the flow of work and maintains focus on production quality products.

4. *Continuous improvement*: continuously improving the effectiveness of work teams by making all processes visible, maximizing effort and ensuring that everyone is committed to improving the manufacturing system.

Key point

Kanban boards and charts provide an overview of the current state of a manufacturing process. They provide a means of tracking workflow, identifying delays and bottlenecks.

Test your knowledge 7.3

Which of the following statements are true?

1. Kanban can assist with the scheduling of workflow.
2. Kanban is more about planning than workflow management.
3. Kanban is an alternative to using statistical process control (SPC).
4. Kanban is only appropriate for use in the manufacturing sector.
5. Kanban is well suited for use in a just-in-time (JIT) context.

Kanban is quite easy to apply. In its simplest form it can be based on a conventional white-board and sticky notes. As the workflow progresses, the sticky notes can be moved into the different 'swim lanes' marked on the board. It is then easy to spot queues and bottlenecks and also easy to reallocate work where necessary. Kanban can also be applied using dedicated software packages (often part of a project management suite). Because the software is network- or cloud-based, it can be accessed, viewed and modified using desktop and factory workstations, as well as laptops, tablets and even mobile phones.

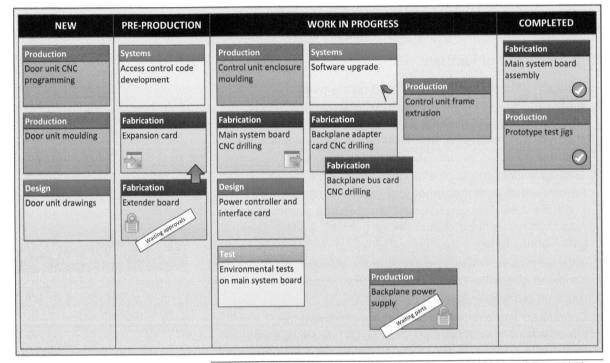

Figure 7.2 The main advantage of the kanban system is being able to quickly and easily visualize processes and the resources allocated to them.

Test your knowledge 7.4

Refer to the kanban chart shown in Figure 7.2 above. Which processes:

a) have been completed?
b) are in the pre-production stage?
c) are in progress but delayed?
d) involve CNC drilling?
e) involve software rather than hardware?
f) have been allocated to the Design Team?

Learning outcome 7.2

Recognize the need for continuous improvement to ensure organizational competitiveness

Having briefly explained the key terms associated with lean manufacturing it is worth taking a look at continuous improvement and the advantages of introducing initiatives such as kaizen and kanban. The benefits of continuous improvement can include:

- increased productivity
- better quality products
- reduced costs
- improved customer satisfaction
- improved morale and increased employee retention
- reduced setup and cycle times
- reduced scrap and rework
- smoother production flow
- reduced inventory of raw materials, work-in-progress and finished goods
- higher worker participation
- a workforce that is more flexible and adaptable with a greater range of skills
- reduction in space and resources
- improved relationships with customers and suppliers
- improved safety.

This is quite a long list and it shows just how important continuous improvement initiatives can be in making a company more efficient and thus more competitive. Competition is something that all engineering companies experience and the activities of competitors are something that will be kept under continuous review.

A key factor in continuous improvement is that it demands the active involvement of everyone in a company. Without a high level of personal commitment to quality improvement, initiatives can often fail to deliver the expected improvements.

Key point

In a business that's committed to continuous improvement, every worker is responsible for delivering quality, not just those with specific responsibility for quality control.

Case study: Luminex plc

Luminex plc is a company that specializes in the manufacture of lighting for use in schools, colleges, airports, theatres and commercial premises. Sensing increasing levels of competition from Eastern Europe, Luminex have introduced a number of continuous improvement initiatives driven by section heads and members of the company's proactive Quality Team. These initiatives have involved staff from all areas in the company as well as all of those directly involved with production activities.

The immediate benefits have been a significant decrease in waste coupled with increased productivity. There have also been less returns of faulty goods from customers. Recent surveys have also indicated an increase in customer satisfaction resulting from reduced delivery times. A new range of studio and theatre lighting products have been well received and contracts have been won against competition from several leading European lighting manufacturers.

Test your knowledge 7.5

List **three** benefits of the continuous improvement initiatives introduced by Luminex. Are these what you would have expected and what further benefits might the company gain?

Learning outcome 7.3

Recognize the importance of improving productivity

What is production?

We use the term 'production' in a wide variety of different contexts so it is worth taking time to clarify what we actually mean by it. In many cases, we take production to mean the output from a manufacturing business but to be meaningful it needs to be stated over a given period of time. For example, the total production of a gas turbine manufacturer might amount to 150 units per quarter or 600 units per annum. However, in practice, the production might not remain constant with time. For example, the annual total of 600 units might have been achieved with 150 units in the first quarter, 200 in the second, 100 units in the third and 150 units in the fourth quarter.

Activity 7.1

The production of basic metals (by value) in the UK for the six-year period from 2008 to 2013 is shown in Table 7.1. Analyse the data by plotting it as a graph and answer the following questions:

1. In what year was the production value greatest?
2. What year had the least production? Suggest a reason for this.
3. What is the average production value for the six-year period?

Table 7.1 Manufacture of basic metals (by value) in the UK from 2008 to 2013 (see Activity 7.1).

Year	2008	2009	2010	2011	2012	2013
Production value (£1,000s)	9,068,985	6,417,023	7,391,321	8,500,016	7,422,006	6,817,405

What is productivity?

Productivity is simply what you get out in relation to what you put in. Productivity can be calculated by dividing average output over a given period of time by the total costs of labour, materials, overheads and resources consumed in that period. Thus, what you get out is the value of products sold to customers and what you put in is all of the costs attributable to the manufacture of the product concerned, all measured over the same period.

As an example, over a period of one quarter (three months) a manufacturer sells products with a total value of £1.5 million. During that period, labour costs amount to £650k whilst materials and component parts cost £150k respectively. All other overhead costs (including capital costs, energy, transport, etc.) amount to £375k. The overall productivity for this quarter can then be calculated from:

$$\text{Productivity (Q1)} = \frac{£1,500,000}{(£650,000 + £150,000 + £375,000)} = 1.28$$

Now suppose that sales in the next quarter have a value of £1.65 million with costs over the same period amounting to £700k for labour with materials and components totalling £190k and overhead costs remaining at £375k. The overall productivity for the next quarter is calculated as follows:

$$\text{Productivity (Q2)} = \frac{£1,650,000}{(£700,000 + £190,000 + £375,000)} = 1.30$$

This indicates an increase in productivity in Q2 when compared with Q1.

An alternative to measuring productivity is to determine the profitability of the operation. For the first quarter this amounts to:

Profitability (Q1) = £1,500,000 – (£650,000 + £150,000 + £375,000) = £325,000

For the second quarter the profitability is:

Profitability (Q2) = £1,650,000 – (£700,000 + £190,000 + £375,000) = £385,000

Activity 7.2

Sales of steel tubes and pipes in the UK during the period 2010 to 2013 are shown in Table 7.2.

1. By how much, expressed as a percentage, has the total production decreased over the four-year period?
2. Determine the value per kg of steel for each year. Has the value per kg increased or decreased over the period?
3. If the downward trend continues at the same average rate per annum, at what year will the production fall to half the amount in 2010?

Table 7.2 Sales of steel tubes and pipes of non-circular cross-section during the period 2010 to 2013 (see Activity 7.2).

Year	2010	2011	2012	2013
Total production (kg)	67,053,599	63,064,224	62,941,419	51,265,209
Value (£1,000s)	141,477	164,157	164,978	133,516

Benefits of improved productivity

Improved productivity is a considerable benefit not only to the company and its employees, shareholders and stakeholders but it is also benefits the region, and the country in which it is located. Profitability contributes to the Gross Domestic Product (GDP) of a country and has a positive effect on earnings, pension security, safety and working conditions. Improvements in productivity result from increased efficiency which, in turn, can be instrumental in reducing waste. Thus, as productivity is increased, there is less waste and consequently less impact on the environment.

It is also important for companies to improve productivity because national and global marketplaces are largely driven by the need to be competitive. This is something that companies of all sizes need to address including large multinationals, national and regional businesses, small to medium enterprises (SME), and even small traders.

Test your knowledge 7.6

Explain why productivity is important. How can productivity be increased?

Test your knowledge 7.7

How is productivity measured? Explain your answer with an example.

Activity 7.3

Use internet research to obtain GDP data for the UK over the past ten years. Plot this data graphically, and use it to answer the following questions.

1. In what year was the maximum GDP achieved?
2. In what year was the GDP a minimum?
3. What trends in GDP can be observed and what does this indicate?
4. Using the GDP for 2010 as a base, what was the average percentage growth in GDP over the next five years?

Learning outcome 7.4

Recognize how to manage the production process

In Section 7.2 we looked at continuous improvement and the advantages that it offers. In this section we look at ways in which continuous improvement can be implemented. To be effective, continuous improvement involves everyone within an organization. It requires leadership from senior management as well as commitment from workers at all levels. Implementation is often carried out within sections and departments and may be led by a Quality Manager and Quality Teams. Since they both lead to improved efficiency, flexible working and multi-skilling are often identified as key areas for continuous improvement.

Plan-Do-Check-Act

One of the most commonly used techniques for continuous improvement is known as the Plan-Do-Check-Act (PDCA) cycle, see Figure 7.3. This is often applied as a model for continuous improvement and it is particularly useful when implementing change, starting a new project, or when improving the design of a process, product or service.

The four basic steps in the cycle of continuous improvement are shown in Figure 7.3.

1. *Plan*: recognize the need for change and define the problem by making observations, collecting data and determining the root cause.
2. *Do*: develop and implement a solution; decide on how you will measure and assess the effectiveness of any changes that are made.
3. *Check*: collect and review the test data. Ensure that the data is valid and evaluate the effectiveness of the changes made.
4. *Act*: implement the changes, embedding them within the process. Update relevant documentation and ensure that all those involved are informed of the changes made.

Finally, use what you have learned to plan further improvements and go round the cycle again.

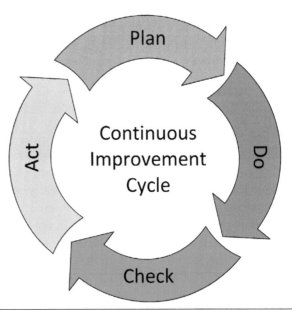

Figure 7.3 The four basic steps in the cycle of continuous improvement.

In Chapter 5 we introduced the production process and the importance of production planning and scheduling. Production planning takes into account the scheduling of materials and parts as well as the availability and loading on production equipment and the flow from product initiation to product delivery. The layout of the production area can have a considerable impact on the flow of product through the different processes involved. The aim should always be that of minimizing the handling of a product so that it flows smoothly from one process to the next. Scheduling is also important as it can reduce lead times and help avoid bottlenecks due to overloading of processes. The four stages of the PDCA cycle can be very useful as a means of optimizing the scheduling process.

Reducing lead times

Lead time is the time that it takes to have stocks of materials and component parts ready for use in a manufacturing processes. With lean (JIT) manufacturing, lead times need to be as short as possible. Most companies achieve a reduction in lead times by one or more methods, including:

- working with suppliers to ensure that they hold materials and parts in readiness so that they can be available quickly as and when required. Close liaison and a good working relationship with suppliers is an essential part of this. By guaranteeing ongoing business, this relationship can also benefit the supplier as well as the manufacturer
- minimizing the number of parts used and ensuring that those used are standard, commonly available parts from a number of different suppliers. Parts that are difficult to source should always be avoided
- using pre-fabricated parts and assemblies where appropriate. This can help reduce assembly times
- using interchangeable parts and components wherever possible. This ensures that parts used in a product can also be used to manufacture a different product
- eliminating unnecessary stages in the production process and streamlining workflow routes so that unnecessary movement of product is avoided.

By reducing lead times delivery schedules will be more accurate. This can reflect positively on a company's brand image. A reduction in lead times can also provide a company with a competitive edge. If a company can offer significantly shorter and more reliable delivery times than its competitors, it will often be able to increase market share and at the same time may even be able to charge more for its products and services.

Layout of plant and equipment

The layout of plant and equipment is also important, particularly when there are significant differences in the design and nature of the product being manufactured. For example, the layout of a press shop used to form small vehicle bodywork components would be unsuitable for forming long sections of steel trunking used for electrical cables. Similarly, the layout of a flow-soldering plant would be ideal for high-volume production of mobile phones but unsuitable for use in the manufacture of control panels for use in a power station.

Figure 7.4 The layout of this large boiler production plant has been designed to accommodate boilers of different types and sizes.

Test your knowledge 7.8

Brief explain the four stages that make up the Plan-Do-Check-Act (PDCA) cycle of continuous improvement.

Test your knowledge 7.9

List **three** advantages of reducing production lead times.

Learning outcome 7.5

Recognize the importance of teamwork and the individual's contribution to effective teamwork

Whether we are aware of it or not, we all work as a member of a team from time to time. A team is simply a group of individuals with a common goal or shared objectives and, most importantly, a team can usually achieve far more than individuals working on their own.

Developing teams

Teams don't just happen – they need to be built with care. An effective team has people in it with a variety of different skills and talent. A team comprising only people with the ability to lead is unlikely to actually get anything done. On the other hand, a team comprising only people with thinking skills is unlikely to be able to do very much practically. What's needed, therefore, is to have a range of balanced skills within the team. We might wish to identify these skills using labels such as 'leaders', 'doers', 'thinkers', 'communicators', 'carers', etc. The team will be able to function effectively as long as we have some of each of these skills present.

Team-building initiatives

Team-building initiatives are frequently used by companies to build and develop their teams. Such initiatives make teams more effective by:

- identifying and reinforcing the individual roles within the team
- clarifying and sharing goals and objectives (to be effective, goals and objectives need to be 'owned' by the team)
- filling any gaps that exist in the team and identifying the means to fill them
- improving communication within the team
- fostering a spirit of trust amongst team members
- recognizing the importance of effective leadership
- emphasizing the need for everyone to participate and, as a consequence, valuing the contribution of all team members.

Team-building activities frequently involve taking team members outside their normal comfort zone, presenting them with entirely

Figure 7.5 Team-building activities frequently involve taking team members outside their normal comfort zone.

new and different challenges. Team-building activities are usually designed so that individual team members can contribute their own skills to the solution of a common problem. Only by working effectively as a team, can the problem be solved.

Communication

In any team effective communication is essential. Without it, the effectiveness of a team can be severely limited. Communication can be enhanced in various ways including:

* locating team members in the same physical area (office, workshop, base room, etc.)
* holding regular team meetings and ensuring that everyone attends them
* using collaborative software applications that can be accessed and used by all members of a team, wherever they happen to be
* using shared diaries and project/process management tools, such as kanban.

> **Key point**
>
> If we only have leaders in a team there will be nobody to do the thinking, working, communicating and caring.

Activity 7.4

You have been asked to identify potential roles for six members of a project team. Which of the team roles (leader, thinker, doer, communicator and carer) are most likely to be appropriate for each of the people listed in Table 7.3.

Table 7.3 Personal qualities of team members (see Activity 7.4)

Team member	Personal qualities
Chris	Calm, thoughtful and focussed. Carefully follows directions
Jack	Good attention to detail. Prefers to take a back seat whenever possible
Mary	Shows initiative. Clear thinker. Good communicator
Rani	Gets on easily with people. Responsive to the needs of others
Rose	Lively and hard-working but easily distracted
Steve	Keen and conscientious but lacking in self-confidence

Test your knowledge 7.10

Describe **three** ways in which communication can be improved between the members of a team.

Review questions

1. Explain what is meant by 'lean manufacture'.

2. Briefly explain the advantages of 'lean manufacture'.

3. Explain how production is measured in a typical manufacturing company.

4. Explain why improvements in productivity can be beneficial to the local environment.

5. Explain the meaning of the term 'Gross Domestic Product'.

6. List **three** different ways of reducing lead times.

7. Explain why production scheduling is important within the production process.

8. Briefly describe **three** different roles that can be performed by members of a team.

9. Explain why balancing the skills of team members is important.

10. Suggest **three** ways of enhancing communication within a team.

Chapter checklist

Learning outcome	Page number
7.1 Explain the meaning of the terms lean manufacture, kaizen, just-in-time and kanban and their overall advantages	166
7.2 Recognize the need for continuous improvement to ensure organizational competitiveness	170
7.3 Recognize the importance of improving productivity	172
7.4 Recognize how to manage the production process	175
7.5 Recognize the importance of teamwork and the individual's contribution to effective teamwork	178

Personal rights and responsibilities

Chapter summary

An organization is only as a good as the people in it. As an employee within an engineering organization you will have a number of rights and also a number of responsibilities. These should all have been made clear to you at the point at which you were first employed; however, some of them might not have seemed very important when you first met them and others might have been difficult to understand. On the other hand, if you have not yet found employment, knowing your rights and also knowing what's expected of you can be extremely important!

This final chapter looks at your rights and responsibilities in some detail. It starts by introducing relevant employment legislation and documentation and then examines the opportunities for progression that exist within most engineering companies. We also look at the support given by a number of representative bodies as well as the benefits that organizations derive from achieving recognition as Investors in People.

Learning outcome 8.1

Identify the relevant organizational documentation and employment legislation in relation to personal rights and responsibilities

Employers are required by law to provide their staff with a written statement of their terms and conditions of employment. In addition, there are several other important documents that organizations should provide so that all staff are made fully aware of the company's procedures and policies. These documents set down the rights of employees as well as those of the employer and they include contracts of employment, employee handbooks, and human resources (HR) policies and procedures.

Contracts of employment

Contracts of employment should not only comply with current employment legislation but should also match the requirements of the organization. Contracts of employment must also be clear regarding the terms and conditions of employment.

Employment Rights Act

The Employment Rights Act 1996 is an Act of Parliament which sets out the framework of employment law in the UK. The legislation is primarily concerned with the rights of the employee and it covers such matters as reasonable notice before fair dismissal, time off for parenting, redundancy and unfair dismissal. The Act attempts to balance the requirements of the job with the fair treatment of those that are doing it. The exact employment rights will vary depending on the kind of job that is being done but in all cases employees have statutory rights as well as those that form the basis of their contracts of employment.

Your statutory rights

Statutory rights are the legal rights that workers are entitled to. There are a few exceptions but, as an employee, you will generally be entitled to these rights as soon as you begin work. They include provisions such as:

- you must be paid at least the National Minimum Wage for your age group
- your employer must not make illegal deductions from your pay
- you must receive a payslip which itemizes your wage and any deductions
- within two months of starting work you must receive a written statement explaining the main terms and conditions of your job
- you have the right to a certain amount of paid holiday each year
- you have the right to take unpaid time off to attend trade union activities, or for study or training if you are aged under 18.

In addition, under health and safety laws, you must be granted daily and weekly rest breaks, and you cannot ordinarily be forced to work more than 48 hours a week.

Contractual rights

You may also have some further rights at work that form part of your contract of employment. These rights will be in addition to your statutory rights and are granted by your employers. Once these rights have been agreed upon, your employer must abide by them. If they don't abide by them, they could be held liable for breach of contract.

Pay rights

For many people the most immediate concern when starting a new job relates to payment for the work done. Pay rights cover issues

Figure 8.1 Contracts of employment are important documents that must comply with current employment legislation.

such as how much and how often you are paid, how you receive your wage, and other important issues.

Employee handbooks

Employee handbooks (or staff handbooks) are not actually a legal requirement but they can be a very useful way of communicating policies and procedures to employees. They are frequently used as a means of supporting and clarifying contracts of employment and they demonstrate good practice in managing human resources. Essential areas of employment practice are outlined in the employee handbook. This not only ensures that staff are aware of their own rights as employees but also makes them aware of the rights the employer.

The primary aim of an employee handbook should be to ensure that there is no possible misunderstanding of the rules that are prevalent in the workplace. For example, an employer might require that all employees working air-side in an airport wear high-visibility jackets or that personnel log their working hours on arrival and departure from their place of employment. In many small organizations an employee handbook can be sufficient to ensure that staff are aware of their own rights as well as those of the employer. Larger companies often have several further policy documents available covering topics such as sickness absence, disciplinary procedures and parental leave.

Key point

Businesses and other organizations have documented procedures and policies relating to employment. These documents include contracts of employment, employee handbooks, and HR policies and procedures.

Test your knowledge 8.1

List **four** statutory employment rights.

Test your knowledge 8.2

Explain the difference between statutory and contractual employment rights.

Working Time Regulations 1998

The Working Time Regulations 1998 implement the European Working Time Directive in the UK. The Working Time Regulations govern the hours that most workers can work and they set limits on an average working week. They also provide for a statutory entitlement to paid leave for most workers, limits on the normal hours of night work and regular health assessments, and include regulations designed to protect young workers.

The Working Time Regulations determine the maximum weekly working time, patterns of work and holidays, plus the daily and weekly rest periods. They also cover the health and working hours of night workers. The regulations apply to both part-time and full-time workers, including the majority of agency workers and freelance staff, although certain categories of workers are excluded.

The Working Time Regulations provide rights that limit work to an average 48 hours a week for most workers. However, individual workers may choose to work longer hours by 'opting out'. The regulations also provide an entitlement for workers to be given paid holiday leave of 5.6 weeks (equivalent to 28 days per annum for someone working full-time, five days a week), as well as a consecutive period of rest of 11 in any 24-hour period, and a statutory 20-minute rest break if the working day is longer than six hours. In addition, employees must have a minimum of one day off each week, and normal working hours of night workers should be limited to an average of eight hours in any 24-hour period. Night workers also have an entitlement to receive regular health assessments.

The regulations for young workers restrict their working hours to 8 hours per day and 40 hours per week. In addition, the rest break is 30 minutes if their work lasts more than 4.5 hours. Young workers are also entitled to two days off each week.

Key point

Employee handbooks are an example of good HR practice. Employee handbooks support contracts of employment and make clear what is required from employees. They also explain company HR policies and identify the provisions for supporting their employees.

Key point

Under the Working Time Directive, you cannot normally work an average of more than 48 hours a week unless you have chosen to 'opt out'. If you are under 18 you can't work more than eight hours a day, or 40 hours a week.

Overtime

Overtime is normally hours that are worked above the normal full-time hours; normal working hours are the hours specified in the terms of employment. Overtime can be voluntary or compulsory but the latter would form part of the terms and conditions of employment. Note that there is no legal right to be paid extra for any overtime worked but this will often be detailed in the contract and terms of employment.

Test your knowledge 8.3

Explain the key provisions of the Working Time Directive.

Personnel records

All organizations, large and small, keep detailed records of their employees. Some of these records are required by law and some are required purely for company purposes.

The records that are required to meet statutory obligations include keeping records of hours worked (under the Working Time Regulations) and pay rates (under the Minimum Wage Act). There is also a need to keep records for tax and National Insurance purposes.

A minimum employee record would include the following key data:

- the full name of the employee
- the full address of the employee
- the date of birth of the employee
- the current job title of the employee (with provision for this to change over time)
- the current pay rate for the employee
- the means of payment (e.g. the employee's bank account number and sort code)
- the employee's tax reference and National Insurance number.

Data needed by the company (rather than for statutory purposes) might include:

- qualifications held
- details of previous employment
- details of education and training.

This information can be invaluable for a number of purposes including planning, appraisal, pay reviews, and the formulation of company training plans.

Activity 8.1

Download a copy of the ACAS booklet on Personnel Data and Record Keeping (available for download from the ACAS website at www.acas.org.uk). Use it to answer the following questions:

1. Explain why personnel records are important (give at least **four** different reasons).
2. List at least **five** key areas for which records should be kept.
3. Explain how good personnel records can help with issues relating to equal opportunities.
4. What key items should be included in a 'person specification'?

Grievance procedures

Grievances are concerns, problems or complaints that employees raise with their employers. Anybody working in an organization may, at some time, have problems or concerns about their work, working conditions or relationships with colleagues that they wish to talk about with management. They want the grievance to be addressed, and if possible, resolved. It is also clearly in management's interests to resolve problems before they can develop into major difficulties for all concerned.

Issues that may cause grievances include:

- terms and conditions of employment
- health and safety
- work relations
- bullying and harassment
- new working practices
- working environment
- organizational change
- discrimination.

Grievances may occur at all levels but a clearly written procedure can help clarify the process and ensure that employees are aware of their rights (such as to have a companion present at a grievance meeting). Some organizations use external mediators to help resolve grievances. Where this is the case the grievance should explain how mediators are used. Employees might also raise issues about matters not entirely within the control of their employee (such as client or customer relationships when an employee is working on another employer's site). These concerns should be treated in the same way as grievances within the organization, with the employer/manager investigating as far as possible and taking action if required.

Appraisals

A performance appraisal (or performance *evaluation*) is a key tool that focusses on an individual's job performance. Appraisals are often carried out on an annual basis and take the form of an in-depth meeting between an employee and his or her line manager. It is important to remember that appraisal is a two-way process and both employee and line manager have a great deal to gain from it. Appraisal should not be rushed and plenty of time should be allowed for discussion. It is also important to prepare for an appraisal in advance.

When preparing for an appraisal you will need to form an objective view of how you have performed in the job role. For example, in what areas could you have improved and in which areas have you exceeded expectation? Your line manager will give you his or her perception of your performance and may suggest ways in which they feel that it could be improved. If this does not match your views, you should be open-minded and accept it as constructive comment rather than criticism. If there was anything wrong with your performance or the way that you have done your job this should have been flagged up already and dealt with by your line manager.

You should also take the opportunity to suggest ways in which your job could be improved, made easier or made safer. In other words, ways in which you could be helped to be more productive. This might highlight a training need or the need for better tools and equipment. Appraisal is your opportunity to raise concerns with your line manager and you should not feel afraid to raise them.

During the appraisal meeting it is important to keep focussed. Because the meeting will normally be held in a private setting you can speak candidly to your appraiser but you do need to stay centred on the matter at hand, namely how your line manager thinks you have been doing and how well he or she thinks you have been doing your job. It is also useful to seek common ground during the meeting.

Outright disagreement with your supervisor rarely works out well for the person being appraised and it can also reflect badly on the appraiser. Nonetheless, it is perfectly fine to take an opposing point of view, especially if you think you have done well in some aspect that your line manager is unaware of. Where a difference of opinion exists, find a compromise solution rather than digging yourself into a position which makes you come across as inflexible.

Appraisals provide you and your line manager with an opportunity for reflection but they are also important in establishing your needs and future development. Don't be afraid to suggest ways in which your personal development can be supported. If there are weaknesses in your performance as well as strengths, then think about how you might improve, perhaps by undergoing training or finding someone to act as a mentor.

Figure 8.2 Appraisals provide you and your line manager with an opportunity for reflection but they are also important in establishing your needs and future development.

Your own personal development

At this point it is well worth thinking about your own personal development and, in particular, developing an outline plan for your own academic and career development. To do this you will need to reflect on your progress and achievements, identifying the knowledge and skills that you wish to develop and improve on. Your own personal development plan will help you to:

- identify gaps in your knowledge and skills
- locate the resources and support that you need
- make the most of opportunities that present themselves
- set realistic goals and targets for your own development
- identify training and areas for specific development
- improve your employability, preparing you for the next step on your career ladder.

Finally, you should always feel free to discuss your personal development with your supervisor or line manager. Appraisal provides you with an excellent opportunity to focus on your needs and aspirations.

Disciplinary procedures

A disciplinary procedure is the means by which rules are observed and standards are maintained. In other words, dealing with all aspects of misconduct in the workplace. Disciplinary procedures are used to help and encourage employees to improve their conduct rather than a means of imposing sanctions on poor performance (which should be dealt with differently). Disciplinary procedures should be fair, effective and consistently applied. Disciplinary situations include misconduct and/or poor performance. Typical situations in which disciplinary procedures are used include wilfully ignoring safety regulations, smoking in prohibited areas, and abuse of company privileges and resources.

Fairness and right of appeal

When drawing up and applying disciplinary procedures, employers should always bear in mind principles of fairness. For example, employees should be fully informed of the allegations against them, together with the supporting evidence, well in advance of any disciplinary meeting. This provides an opportunity for employees to challenge allegations made against them before any decisions are reached. Employees should also be given the right to appeal against any decisions that are made.

A disciplinary procedure should be written and made available to all employees. It should:

- be non-discriminatory
- provide for matters to be dealt with speedily
- allow for information to be kept confidential
- tell employees what disciplinary action might be taken
- say what levels of management have the authority to take the various forms of disciplinary action
- require employees to be informed of the complaints against them and supporting evidence before a disciplinary meeting
- give employees a chance to have their say before management reaches a decision
- provide employees with the right to be accompanied
- provide that no employee is dismissed for a first breach of discipline, except in cases of gross misconduct.

Activity 8.2

Download a copy of the ACAS booklet on Discipline and Grievances at Work (available for download from the ACAS website at www.acas.org.uk. Use it to answer the following questions:

1. Explain how a mediator can sometimes help resolve disciplinary or grievance issues.

2. List **four** circumstances in which mediation is inappropriate.

3. Explain why rules and performance standards are important.

4. Briefly explain the role of a 'companion' at a grievance meeting.

Key point

Problems and concerns are often raised and settled as a matter of course in organizations where managers have an open policy for communication and consultation. In such cases, problems can be quickly raised and settled during the course of everyday working relationships.

Employment legislation

In Chapter 1 we mentioned some of the legislation that employers and their employees need to be aware of, notably the Health and Safety at Work Act. In addition to this important legislation there are several other UK Acts of Parliament that you need to be aware of. They include:

* The Data Protection Act
* Equal Opportunities Policies
* The Sex Discrimination Act
* The Race Relations Act
* The Race Relations (Amendment) Act
* The Human Rights Act
* The Disability Discrimination Act.

We will take a brief look at this legislation, starting with a brief recap of The Health and Safety at Work Act.

The Health and Safety at Work Act

The Health and Safety at Work Act seeks to promote greater personal involvement in health and safety with a particular emphasis on individual responsibility and accountability. The Act applies to *people*, not to premises and it covers all employees in all employment situations. The Act also requires employers to take account of the fact that other persons, not just those that are directly employed, may be affected by work activities. It also places certain obligations on those who manufacture, design, import or supply articles or materials for use at work to ensure that these can be used safely and do not constitute a risk to health.

The Data Protection Act

The Data Protection Act controls how your personal information is used by organizations, businesses or the government. The Act makes everyone responsible for collecting and using data follow strict rules called 'data protection principles'. In particular, they must make sure the information is:

- accurate
- kept safe and secure
- used fairly and lawfully
- used only for limited, specifically stated purposes
- used in a way that is adequate, relevant and not excessive
- kept for no longer than is absolutely necessary
- handled according to people's data protection rights
- not transferred outside the European Economic Area without adequate protection.

Activity 8.3

Visit the UK Government website (www.gov.uk) and search for information on the Data Protection Act. Use it to answer the following questions:

1. Name **four** types of information where there is stronger legal protection under the Act.
2. How can you find out what information a particular organization holds about you?
3. Name **three** types of information that an organization can withhold.
4. Can an organization charge for supplying the information?
5. What can you do if you think that your data has been misused by an organization?

The Equality Act 2010

The Equality Act 2010 legally protects people from discrimination in the workplace and in wider society. It has replaced previous anti-discrimination laws with a single Act, making the law easier to understand and strengthening protection in some situations. It sets out the different ways in which it is unlawful to treat someone. Before the Equality Act came into force there were several pieces of legislation to cover discrimination, including:

- The Sex Discrimination Act 1975
- The Race Relations Act 1976
- The Disability Discrimination Act 1995.

What do we mean by discrimination?

A person discriminates against another person if, because of a *protected characteristic*, the person treats the other less favourably than he or she would treat others. Protected characteristics include:

* age
* disability
* gender reassignment
* marriage and civil partnership
* pregnancy and maternity
* race (colour, nationality, ethnic or national origins)
* religion or belief
* sex (gender)
* sexual orientation.

Here are five, very different situations that are in contravention of the Equality Act:

1. Bob is aged 62 and has applied for a job as a long-distance lorry driver. He has the necessary HGV certification and a clean driving licence. He is also in good health but, on application, he was told that the reason he could not be considered is that he is too old for the job.
2. Assif is the only Muslim in a small team of security officers. After a terrorist attack in the UK, his colleagues begin to treat him differently. They stop talking to him and no longer invite him to join them in workplace social events. At a team meeting, his line manager jokingly suggests that Assif's family must have links with a terrorist group. Dismayed by this assertion, Assif feels that he has become a 'pariah' so he resigns.
3. Gordon is a heterosexual man who is HIV-positive. He told his employer when he joined the company of his HIV-positive status, but wanted this information to be kept from his colleagues. His employer treated this information in confidence, but a colleague found out and told other employees. Gordon overheard one colleague saying that his illness was a punishment from God for being gay (although he is, in fact, heterosexual). Several colleagues went on to express their fear of being infected and the employer was eventually persuaded to dismiss Gordon.
4. Jane, who is 29 and has one small child, applied for a post with a prospective employer. She was asked at the first interview stage if she intends to have any more children. Although she felt embarrassed by and uncomfortable with the question, she said that it was not her intention to have any more children. Jane didn't get the job and she felt that the questions had been aimed at eliminating women of child-bearing age who might later

Key point

The Equality Act 2010 legally protects people from discrimination in the workplace and in wider society. It replaces previous anti-discrimination laws such as the Sex Discrimination Act, Race Relations Act, and Disability Discrimination Act.

become pregnant. She then found out that two men in their 40s had been called back for second interviews.

5. Marion was recently appointed as a production worker. She is a lesbian and has recently been subjected to homophobic taunts and abuse by other workers. This has mainly taken the form of verbal abuse using words such as 'dyke'. Marion reported these incidents to her line manager but was told that such taunts were an accepted part of the job and she had to 'grin and bear it' or she was 'in the wrong job'. The abuse continued and, as a result of stress, Marion took sick leave and never returned to work.

Test your knowledge 8.4

Select any **three** of the scenarios listed and for each explain why the Equality Act has not been complied with. In each case, suggest which protected characteristics are involved and what subsequent action should be taken.

Learning outcome 8.2

Identify the personal opportunities for development and progression

Throughout your career as an engineer you will have various development and progression opportunities. In order to take full advantage of them you need to be aware of what they are and how they can work for you. It is also useful to keep in mind the different ways that your career might develop and how you can be best placed to seize opportunities when they arise. This means keeping an eye on the career paths that are open to you and knowing how you can best prepare yourself for your next move, as well as the ones that might come later. In all of this it is important to keep an open mind by not ruling things out at an early stage. By keeping your options open you will have the widest possible range of opportunities. Now let's take a look at some of them.

Company training programmes

All companies provide training programmes for their employees. These are designed primarily to meet the needs of employers but they are also of benefit to employees. Larger companies are able to offer a greater range of company training than smaller firms. Typical examples of company training include first aid training, fire training, etc. Note that in some cases your employment might be conditional on you receiving company training and being able to demonstrate competence in the skills that you have learned. For example, if you need to use an abrasive wheel as part of your job you might receive specific training relating to the use of abrasive wheels followed by some form of practical competence assessment.

Apprenticeships

Apprenticeships are normally provided for young people entering employment. They are structured programmes and are designed to provide the vocational skills needed for working within a specific industry. Apprenticeships mix on-the-job training with classroom learning. They provide you with the skills you need for your chosen career that will also lead to nationally recognized qualifications. As an apprentice you earn while you learn and receive other benefits as well.

Figure 8.3 Apprenticeships combine on-the-job training with classroom learning and provide the skills you need for your chosen career as well as a nationally recognized qualification.

Organizational training opportunities

Companies view training as a way of improving the effectiveness of their current workforce. Programmes are often advertised internally and employees may elect to join them or they might be encouraged by line managers, mentors and colleagues. They might involve release from employment to permit attendance or employees might be expected to give up some of their own time to attend. However, in many cases there is some give and take concerning the time needed for training. The benefits of organizational training include:

- improved productivity and adherence to quality standards
- employees develop skill-sets that allow them to work more flexibly and undertake a greater variety of work
- improved ability to achieve goals that form part of a company's strategic plan
- increased ability to respond effectively to change.

Delivery of training

Training can be delivered in various ways, including:

- on-the-job learning (delivered internally or by external providers)
- off-the-job training (delivered by external providers)
- in-house mentoring schemes
- in-house training (part of company training programmes)
- individual self-study
- day release or block release training (delivered by a college or training centre).

Other professional development opportunities

Other professional development opportunities include *promotion*, *transfer* to another area within the company, *higher education* and *professional qualifications*. Most, if not all, of these opportunities are likely to arise from time to time and form part of your appraisal. They should be viewed as working together to make you into a competent person with a range of skills and abilities. Often one opportunity will follow another. For example, higher education followed by a professional qualification is likely to lead to promotion.

Test your knowledge 8.5

Explain what is meant by an apprenticeship. What advantages are there of being an apprentice?

Learning outcome 8.3

Identify the representative bodies in the engineering sector that support personnel and organizations

A number of recognized bodies represent the interests of engineers in the UK. They include:

- trade unions

- professional bodies
- employers' organizations
- industry training support.

Trade unions

A trade union is an organized association of people who work in a profession, trade or group of trades. A trade union exists to protect and further the rights and interests of its members. Several of the largest trade unions active in the engineering, manufacturing, production and communications sectors are:

- *CWU*: a trade union for the communications industry in the UK, representing members in postal, telecom, mobile, administrative and financial companies.
- *GMB*: a general trade union to which anyone can belong
- *Prospect*: a trade union representing engineers, scientists, managers and specialists in areas as diverse as agriculture, defence, energy, environment, heritage, shipbuilding, telecoms and transport.
- *Unite*: general trade union representing workers across the country and all industrial sectors (Community section of Unite also covers people who are out of work, e.g. students, volunteers, retired people, people who are unemployed).

Activity 8.4

Visit the TUC website at www.tuc.org.uk and download a copy of *Trade Unions at Work*. Use it to answer the following questions:

1. Name **five** different things that unions do for their members.
2. Explain what is meant by 'union recognition'.
3. List the **five** principal aims of the TUC.
4. Explain what is meant by an 'employment tribunal'.

Professional bodies

A professional body is an organization of professional members. In some professions membership is compulsory in order to work or *practice*. This usually depends on whether or not the profession requires the professional to have a 'licence to practice', or to be on a professional register in order to do their job. Professional bodies have a number of functions including:

- representing the interests of their members

- setting and assessing professional standards and examinations
- providing support for *continuing professional development* (CPD) through learning opportunities and tools for recording and planning
- publishing professional journals or magazines
- providing *networking opportunities* for professionals to meet and discuss their field of expertise
- issuing *codes of conduct* (or equivalent standards) to guide and inform professional behaviour
- dealing with complaints made against professionals
- providing careers support and opportunities for students, graduates and people already working.

Activity 8.5

Use internet or library research to obtain information about **three** professional bodies that represent the interests of engineers in the UK. State the full name of each body and the range of engineering specialisms that it caters for (see Appendix 4). Also list the grades of membership available along with the experience and qualifications required.

Employers' organizations

An employers' organization is an association that represents the interests of employers within a particular industry sector (or within related industry sectors). Employers' organizations often negotiate with trade unions on a collective basis as well as sharing information and advice of benefit to their members. In the UK, the Engineering Employers' Federation (EEF) seeks to work with all parties in the sector to support and champion manufacturing and engineering in the UK and also in Europe. The organization states that it is 'the voice of UK manufacturing and engineering and a leading provider of business support'. The EEF works with a wide range of people, from industry leaders, managers and professionals, to young people, apprentices, policy-makers and the media.

Industry training support

In addition to employers' associations there are Sector Skills Councils (SSCs) covering specific UK industries. These organizations have four key goals:

- to improve the supply of learning and training within the industry
- to support employers in developing and managing apprenticeships

Key point

A trade union is an organized association of people who work in a profession, trade or group of trades. A trade union exists to protect and further the rights and interests of its members.

- to reduce skills gaps and shortages and improve productivity
- to boost the skills of their sector workforces.

The SSC for engineering and manufacturing is the Science, Engineering, Manufacturing and Technologies Alliance (SEMTA). SEMTA is a not-for-profit organization responsible for engineering skills for the future of the UK's most advanced sectors. SEMTA is committed to inspiring the next generation of engineers, showcasing the best of British engineering talent and driving excellence in science, technology, engineering and mathematics (STEM) teaching.

Test your knowledge 8.6

State the meaning of each of the following abbreviations:

1. TUC
2. CPD
3. SSC
4. EEF
5. SEMTA

Learning outcome 8.4

Explain the implications that 'Investors in People' has on an organization and its personnel

The Investors in People (IIP) standard has been used by UK organizations as a means of business improvement for more than 25 years. The standard sets out a framework of good practice in the management of people. Accredited IIP organizations have been shown to be more profitable, sustainable and optimistic about the future. In a recent survey, 60% of IIP-accredited organizations predicted business growth compared with the UK establishment average of 47%. The key principles associated with IIP are:

- *Leading:* creating purpose in a fast-changing environment: high-performing organizations foster leadership skills at every level of the organization in order to deliver outstanding results.
- *Supporting:* moving towards flatter structures to enable faster decision-making, customer focus and agility. Successful organizations reduce their overheads and provide better service for their customers as a result of a supportive approach.
- *Improving*: the best organizations are always looking for opportunities to improve by seeking every marginal gain. They

Key point

A professional body is an organization that represents the interests of its professional members. An important role of all professional bodies is that of facilitating continuing professional development (CPD) and providing careers support for students, graduates and people already working in the industry seeking professional recognition.

Key point

In the UK the Engineering Employers' Federation (EEF) is the leading organization that represents engineering employers whilst the Science, Engineering, Manufacturing and Technologies Alliance (SEMTA) is the leading organization for education and training.

know that every small change adds together to enable them to constantly outperform.

Within the framework there are nine individual themes or 'indicators' and three indicators are associated with each of the key principles. Organizations are able to rate their performance against each of the nine indicators using four levels of performance:

1. *Developed:* in place and understood.
2. *Established:* engaging and activating.
3. *Advanced:* creating positive outcomes.
4. *High Performing:* embedded and always improving.

Steps to IIP accreditation

There are four basic steps to IIP accreditation:

1. *Discovery*: exploring online self-assessment and resources.
2. *Online assessment*: understanding employees' views of the organization.
3. *Employee interviews and observation:* exploring key themes emerging from the online assessment through face-to-face meetings and observations.
4. *Accreditation report:* finding the award level and benchmarking performance, including gaining insights from the online data, interviews and assessment and also comparing performance with best practice in order to identify next steps.

IIP award levels

The IIP award levels include:

- *Accredited:* where all nine indicators are at 'Developed' level
- *Silver:* where all nine indicators are at 'Developed' level and seven of the nine indicators are at 'Established' level
- *Gold:* where all nine indicators are at 'Developed' and 'Established' levels and seven of the nine indicators are at 'Advanced' level
- *Platinum:* where all of the nine indicators are at 'Developed', 'Established' and 'Advanced' levels and seven of the nine indicators are at 'High Performing' level.

Key point

IIP accreditation shows that an organization is committed to excellence. It is a way of demonstrating that an organization recognizes the value of its people and is willing to put into place the necessary changes to achieve leading-edge performance.

Case study: HLS Systems

HLS Systems has recently achieved Gold Investors in People accreditation, putting the company in the top 3% of IIP accredited companies. HLS Systems has held an IIP award since

2007 and the new award comes after a recent comprehensive audit by IIP assessors who evaluate a company's performance and approach to its people through key areas evidenced through interviews with staff at all levels within the organization.

Key areas of strength identified by IIP assessors included HLS Systems' embedded culture of training and mentoring, and the quality of leadership was referred to as 'inspirational' and 'motivational'. The final IIP report states: 'All employees at HLS Systems have a sense of inclusion in planning, communications, innovation and operations. Knowledge and skills are shared and information is freely available in an environment where people value being involved and part of the decision making process.' The report continued by saying: 'People have a strong sense of feeling valued and being appreciated. They feel they have constructive and supportive feedback which helps to develop skills and expertise, as well as being well informed and involved.'

Mark Williams, Managing Director of HLS Systems, commented that: 'HLS is delighted to have been recognized for the hard work and inclusivity within our organization. HLS has always been committed to the development and progression of its staff and being awarded the IIP Gold standard shows how much we value and support our staff with structured training and mentoring. This has helped us retain key staff and improve our business performance over the past five years.'

Activity 8.6

Read the HLS Systems case study and answer the following questions:

1. What proportion of IIP accreditations are made at Gold level?
2. What evidence contributed to the audit by IIP assessors?
3. In what key business areas do all HLS Systems staff feel 'included'?
4. In addition to training what method of professional development is used at HLS Systems?
5. What **two** benefits of working towards IIP accreditation are mentioned?

Test your knowledge 8.7

List the **four** steps to IIP accreditation.

Review questions

1. Identify **three** key organizational documents associated with rights and responsibilities of employees.

2. Explain the purpose of an employee handbook.

3. List **three** important rights that are protected by the Working Time Directive.

4. List **five** key items of information that would be present in an employee's personnel record.

5. Explain the difference between a disciplinary procedure and a grievance procedure.

6. Explain how an individual's training needs are identified by means of appraisal.

7. List **five** representative bodies that support employers and employees within the engineering sector.

8. Explain the implications of 'Investors in People' on an organization and on its employees.

Chapter checklist

Learning outcome	Page number
8.1 Identify the relevant organizational documentation and employment legislation in relation to personal rights and responsibilities	184
8.2 Identify the personal opportunities for development and progression	196
8.3 Identify the representative bodies in the engineering sector that support personnel and organizations	198
8.4 Explain the implications that 'Investors in People' has on an organization and its personnel	201

Appendix 1
Sample assessment

This assessment consists of 40 multiple choice questions. Select one answer to each question. The numbers in square brackets below each question relate to the syllabus sub-group (there is one multiple-choice question for each sub-group). Note that within a real assessment, questions will appear in random sub-group order within each section.

Time allowed: 60 minutes

Section 1

1. Which one of the following individuals is responsible for investigating incidents involving contaminated waste?
   a) HSE Inspectors
   b) Company Safety Officers
   c) Environmental Health Officers
   d) Elected Safety Representatives
   [1.1]

2. Under the Management of Health and Safety at Work Regulations employers have a statutory duty to:
   a) create a designated outdoor area for smokers
   b) provide all personnel with regular health screening
   c) employ a medical team to deal with accidents and emergencies
   d) appoint one or more competent person to oversee health and safety
   [1.1]

3. Which one of the following is **not** normally part of a health and safety policy statement?
   a) Arrangements for monitoring accidents and first aid
   b) Emergency contact numbers for all staff working on-site
   c) Procedures relating to the safe handling and storage of substances
   d) Details of how employees are consulted on matters relating to health and safety
   [1.1]

4. Under which one of the following regulations must a company ensure that dangerous chemicals are clearly marked?
a) Working Time Regulations
b) Manual Handling Operations Regulations
c) Provision and Use of Work Equipment Regulations
d) Control of Substances Hazardous to Health Regulations
[1.1]

5. Under which one of the following regulations must a company ensure that its machines and tools are fit for purpose?
a) Working Time Regulations
b) Manual Handling Operations Regulations
c) Provision and Use of Work Equipment Regulations
d) Control of Substances Hazardous to Health Regulations
[1.1]

6. The Manual Handling Operations Regulations require employers to:
a) provide washing facilities
b) provide hard hats for all site visitors
c) assess tasks that involve lifting heavy objects
d) ensure that staff receive regular fitness assessments
[1.1]

7. Figure A1.1 shows the steps involved in performing a risk assessment. In what order should these steps be carried out?
a) 2, 3, 5, 4, 1
b) 3, 5, 2, 1, 4
c) 4, 2, 5, 3, 1
d) 5, 2, 3, 4, 1
[1.2]

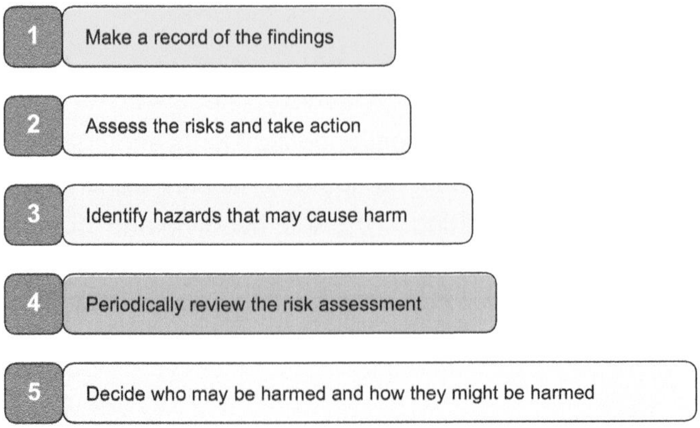

1 Make a record of the findings

2 Assess the risks and take action

3 Identify hazards that may cause harm

4 Periodically review the risk assessment

5 Decide who may be harmed and how they might be harmed

Figure A1.1 The steps involved in performing a risk assessment.

8. Figure A1.2 shows some statements with the words **do** and **don't** missing. Which one of the following is correct?
   a) **Do** should appear in 1, 2, 3; **don't** should appear in 4, 5.
   b) **Do** should appear in 2, 3, 4; **don't** should appear in 1, 5.
   c) **Do** should appear in 3, 4; **don't** should appear in 1, 2, 5.
   d) **Do** should appear in 2, 3, 4, 5; **don't** should appear in 1.
   [1.2]

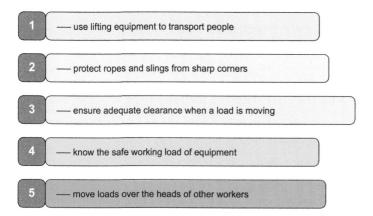

1 —— use lifting equipment to transport people

2 —— protect ropes and slings from sharp corners

3 —— ensure adequate clearance when a load is moving

4 —— know the safe working load of equipment

5 —— move loads over the heads of other workers

Figure A1.2 Statements relating to question 8.

9. Breathing apparatus can be essential when:
   a) working at height
   b) entering a quarantine store
   c) working in a confined space
   d) identifying empty gas cylinders
   [1.2]

10. A permit to work is usually issued and signed off by:
    a) a site manager
    b) senior management at head office
    c) a local environmental health officer
    d) the contractor who is carrying out the work
    [1.2]

11. Figure A1.3 shows some statements with the words **do** and **don't** missing. Which one of the following is correct?
    a) **Do** should appear in 1, 2, 3; **don't** should appear in 4, 5, 6.
    b) **Do** should appear in 1, 2, 5, 6; **don't** should appear in 3, 4.
    c) **Do** should appear in 1, 4, 5, 6; **don't** should appear in 2, 3.
    d) **Do** should appear in 2, 3, 4; **don't** should appear in 1, 5, 6.
    [1.2]

1 — inspect fire and smoke alarms regularly

2 — ignore the advice of a Fire Prevention Officer

3 — allow rubbish to accumulate in the work area

4 — ensure that oily rags and other waster is stored in metal bins

5 — ensure that electrical repairs are carried out by qualified persons

6 — ensure that smoking is banned in areas where flammable materials are stored

Figure A1.3 Statements relating to question 11.

12. Which item of lifting equipment is shown in Figure A1.4?
   a) Platform
   b) Scissor lift
   c) Pallet truck
   d) Forklift truck
   [1.3]

Figure A1.4 Lifting equipment.

13. Which one of the features shown in Figure A1.5 is a shackle?

a) A

b) B

c) C

d) D

[1.3]

Figure A1.5 Which one of these features is a shackle?

14. The ratings marked on the handle of the ratchet winch shown in Figure A1.6 comply with:

a) manufacturers' requirements

b) the Health and Safety at Work Act

c) BS Safe Working Load (SWL) markings

d) Lifting Operations and Lifting Equipment Regulations

[1.3]

Figure A1.6 Ratchet winch.

15. Which one of the following gives the maximum recommended spread on the sling arrangement shown in Figure A1.7?
    a) 45°
    b) 90°
    c) 120°
    d) 180°
    [1.3]

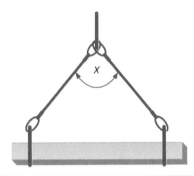

Figure A1.7 Sling arrangement.

16. Figure A1.8 shows some chemicals in a store. Which one of the chemicals is hazardous?
    a) A
    b) B
    c) C
    d) D
    [1.3]

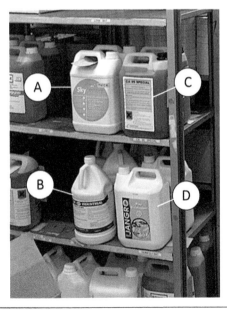

Figure A1.8 Chemicals in a store.

17. Welding and other 'hot work' on tanks and drums that have contained flammable material is controlled under the:
    a) Controlled Waste Regulations
    b) Pollution Prevention and Control Act
    c) Care of Substances Hazardous to Health Regulations
    d) Dangerous Substances and Preparations and Chemical Regulations

    [1.4]

18. The purpose of an environmental management system is to:
    a) minimize harmful effects on the environment
    b) monitor and record the release of harmful waste materials
    c) maximize the use of energy in order to protect the environment
    d) collect and store hazardous materials in tanks and sealed containers

    [1.4]

19. The Energy Technology Product List (ETL) is a list of:
    a) suppliers of high-carbon energy sources
    b) manufacturers of high-technology energy sources
    c) government-approved energy-efficient plant and equipment
    d) alternative energy sources that can be used to replace fossil fuels

    [1.4]

20. Hydroelectric energy involves:
    a) pumping warm water through a heat exchanger
    b) passing a flow of water through a turbine-driven generator
    c) transferring heat energy from a warm place to a cold place
    d) using continuous tidal flow to power a turbine-driven generator

    [1.4]

21. Which one of the following chemical substances causes acid rain when released into the atmosphere?
    a) Carbon dioxide
    b) Copper oxide
    c) Sulphur dioxide
    d) Hydrochloric acid

    [1.4]

Section 2

22. Batch flow production is frequently used for manufacturing:
   a) identical products on a continuous basis
   b) large quantities of identical items for constant demand
   c) small quantities of product to a specific client specification
   b) one-off items where each product needs to be manufactured to a different specification
   [2.1]

23. When selecting a method of production, it is essential to:
   a) have a detailed marketing plan
   b) be aware of market requirements
   c) use only existing parts and materials
   d) employ new personnel as far as possible
   [2.1]

24. Select A and B from the list below to complete the statement:
   'As part of lean manufacturing it is possible to A _____ by B _____.'

Phrase A	Phrase B
1) increase scrap	a) increasing push production
2) increase lead time	b) increasing quality controls
3) reduce costs	c) reducing safety training
4) reduce accidents	d) reducing stockholding

   [2.1]

25. Select A and B from the list below to complete the statement:
   'A work schedule should include A _____ because B _____.'

Phrase A	Phrase B
1) training plans	a) accidents might occur
2) tooling details	b) new staff might be needed
3) first aid information	c) there's a need to reduce waste
4) disciplinary procedures	d) it's required for machining and fitting

   [2.1]

26. Select A and B from the list below to complete the statement:
   'The symbol shown in Figure A1.9 is A _____ used in a B _____.'

Phrase A	Phrase B
1) a process	a) control chart
2) a decision	b) flowchart

3) a start point   c) Gantt chart
4) a terminator   d) conversion chart
[2.1]

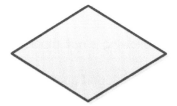

Figure A1.9 Symbol required for question 26.

27. Quality assurance is defined as:
  a) tests and measurements made to ensure that a product is 'fit for purpose'
  b) applying statistical process control to guarantee that products work correctly
  c) the implementation of written procedures to ensure that quality control takes place
  d) checks made during the production process to ensure that a product complies with its published specification
  [2.2]

28. The advantages of adopting the BS EN 9001 quality assurance standard include:
  a) an increase in both cost and efficiency
  b) a reduction in both cost and efficiency
  c) an increase in cost and a reduction in efficiency
  d) a reduction in cost and an increase in efficiency
  [2.2]

29. Inspection is a quality control tool in which:
  a) any deviation from what is expected is detected
  b) the quality of a product is measured and recorded
  c) statistical process control is used to locate faulty parts
  d) test procedures are used to confirm quality assurance standards
  [2.2]

30. Mean Time Between Failures (MTBF) is defined as the:
  a) average time that it takes to repair a component or product when it fails
  b) average amount of time that a component or product functions before failing

c) minimum amount of time that a component or product functions before failing

d) maximum amount of time that a component or product functions before failing

[2.2]

31. Which one of the following is **not** important when developing a quality plan?

a) Agree budgets to support quality activities.

b) Be fully aware of customer and client expectations.

c) Set up systems to measure quality and report progress.

d) Stimulate growth by publishing regular production data.

[2.2]

32. A culture within an organization that seeks to raise the quality in all aspects of the company's operation, with processes being done right the first time and defects and waste eradicated from operations, is referred to as:

a) Total Quality Control

b) Total Quality Assurance

c) Total Quality Management

d) Total Quality Improvement

[2.2]

33. Just-in-time (JIT) manufacturing is sometimes also referred to as:

a) batch production

b) flow production

c) stockless production

d) intermittent production

[2.3]

34. Lean manufacturing offers:

a) a way to reduce production costs but with more waste

b) increased production costs but less need for a skilled workforce

c) reduced space requirements but greater need for a skilled workforce

d) reduced production costs but a need for a large stock of parts and materials

[2.3]

35. Improved productivity can be shown to have:

a) less waste and less impact on the environment

b) less waste and more impact on the environment

c)   more waste and less impact on the environment

d)   more waste and more impact on the environment

[2.3]

36.   The four consecutive steps of the cycle of continuous improvement are:

a)   plan, act, check, do

b)   plan, do, act, check

c)   plan, do, check, act

d)   plan, check, do, act

[2.3]

37.   Which one of the following is **false**?

a)   Effective communication helps make a team work better.

b)   A team is more successful if all of its members are leaders.

c)   A range of skills and abilities can help make a team more effective.

d)   Teams can achieve more than a number of individuals working alone.

[2.3]

38.   Match **each** of the listed organizational documentation to the purposes listed:

Documentation

1)   Disciplinary procedure

2)   Contracts of employment

3)   Grievance procedures

4)   Staff handbook

Purpose

a)   Legally binding agreement between employer and employee

b)   Summarizes company policies and procedures

c)   Defines a process for resolving problems

d)   Code of practice that is used to deal with misconduct

[2.4]

39.   Match **each** of the listed organizational documentation to the purposes listed:

Documentation

1)   Sex Discrimination Act

2)   Data Protection Act

Purpose

a)   Protects people in the workplace

b)   Seeks to eliminate discrimination due to age

3) Equal Opportunities Policies   c) Seeks to eliminate discrimination due to gender

4) Health and Safety at Work Act   d) Protects confidentiality of personal information

[2.4]

40. Match **each** of the listed organizations to the purposes listed:

Organizations	Purpose
1) Trade unions	a) Represents manufacturers and producers
2) Professional bodies	b) Promotes the interests of groups of workers
3) Sector Skills Councils	c) Provides recognized status for its members
4) Employers' associations	d) Responsible for developing training in the industry

[2.4]

Answers

1. c	2. d	3. b	4. d	5. c
6. c	7. b	8. b	9. c	10. a
11. c	12. b	13. d	14. d	15. c
16. c	17. d	18. a	19. c	20. b
21. c	22. c	23. b	24. 3d	25. 2d
26. 2b	27. c	28. d	29. a	30. b
31. d	32. c	33. c	34. c	35. a
36. c	37. b	38. 1d, 2a, 3c, 4b	39. 1c, 2d, 3b, 4a	40. 1b, 2c, 3d, 4a

Appendix 2
Abbreviations

AC	Alternating current
ACOP	Approved codes of practice
APW	Air-pressurized water
BIT	Business improvement techniques
BOM	Bill of materials
BS	British Standard
BSI	British Standards Institution
CAD	Computer aided design
CAM	Computer aided manufacturing
CCL	Climate change levy
CE	Conformité Européene (European Conformity)
CNC	Computer numerical control
CHIP	Chemicals Hazard Information and Packaging for Supply Regulations
CHP	Combined heat and power
CIBSE	Chartered Institution of Building Services Engineers
CIPHE	Chartered Institute of Plumbing and Heating Engineering
COSHH	Control of Substances Hazardous to Health
CPD	Continuing professional development
CPS	Carbon price support
CWR	Controlled Waste Regulations
DC	Direct current
DECC	Department of Energy and Climate Change
DSEAR	Dangerous Substances and Explosive Atmospheres Regulations
EAL	EMTA Awards Ltd
ECA	Enhanced Capital Allowance
EEF	Engineering Employers' Federation
EHO	Environmental Health Officer
EMS	Environmental management system
EMTA	Engineering and Marine Training Authority
EPA	Environmental Protection Act
EPR	Environmental Permitting Regulations
EQA	External quality assurer
ETL	Energy Technology Product List
EU	European Union
FMEA	Failure Mode and Effects Analysis
GDP	Gross Domestic Product

HR	human resources
HSC	Health and Safety Commission
HSE	Health and Safety Executive
IAC	Industry Apprentice Council
ICME	Institute of Cast Metals Engineers
ICT	Information and communications technology
IET	Institution of Engineering and Technology
IIP	Investors in People
IMarEST	Institute of Marine Engineering, Science and Technology
IMechE	Institution of Mechanical Engineers
IOM3	Institute of Materials, Minerals and Mining
IOSH	Institution of Occupational Safety and Health
ISO	International Organization for Standardization
IT	Information technology
JIT	Just-in-time
LA	Local authority
LCL	Lower control limit
LED	Light-emitting diode
LOLER	Lifting Operations and Lifting Equipment Regulations
LSL	Lower specification limits
MHOR	Manual Handling Operations Regulations
MRP	Manufacturing Requirements Planning
MSDS	Material Safety Data Sheets
MTBF	Mean time between failures
MTTF	Mean time to failure
NC	Numeric control
NI	Nuclear Institute
PCB	Printed circuit board
PDCA	Plan-Do-Check-Act
PPCA	Pollution Prevention and Control Act
PPE	Personal protective equipment
PUWER	Provision and Use of Workplace Equipment Regulations
PV	Photovoltaic
QA	Quality assurance
QC	Quality control
QCF	Qualifications and Credit Framework
QMS	Quality Management System
R&D	Research and development
RIDDOR	Reporting of Injuries, Diseases and Dangerous Occurrences Regulations
RSI	Repetitive strain injury
SCM	Supply Chain Management

SEMTA	Science, Engineering, Manufacturing and Technologies Alliance
SOE	Society of Operations Engineers
SOP	Standard operating procedures
SPC	Statistical process control
SSC	Sector Skills Council
STEM	Science, technology, engineering and mathematics
SWL	Safe working load
SWP	Safe working procedures
TQM	Total Quality Management
TSO	Trading Standards Officer
UCL	Upper control limit
UK	United Kingdom
USL	Upper specification limits
UTC	University Technical College
UV	Ultraviolet
VA	Value analysis

Appendix 3
Using the Casio
fx-83 calculator

Figure A3.1 The Casio fx-83 scientific calculator.

The Casio fx-83 calculator is currently widely available and is highly recommended for use by engineering students at Level 2 and Level 3. The calculator has over 250 functions and it incorporates a 'natural display' that supports input and output using mathematical notation, such as fractions, roots, etc.

Initializing the calculator

The keystrokes shown in Figure A3.2 can be used to return the calculator's settings and modes to their initial (default) values. Note that this operation will also clear any data from the calculator's memory. The fx-83's Clear menu is shown in Figure A3.3.

Figure A3.2 Key entry required to initialize the calculator.

Setting the calculator mode

The fx-83 provides three different calculation modes: COMP is used for general calculations, STAT is used for statistical calculations, and TABLE is used to generate a table of values based on a given expression. For Level 3, you will only need to use the COMP mode. The Mode menu is shown in Figure A3.4 and the keystrokes required to set the calculator to the COMP mode are shown in Figure A3.5.

Figure A3.3 The fx-83's Clear menu will allow you to reset and initialize the calculator.

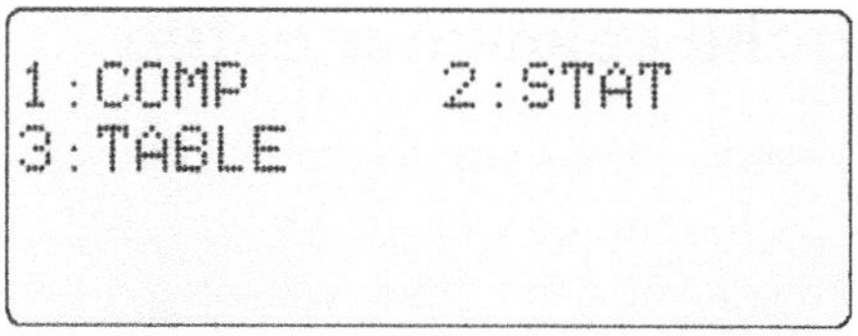

```
1:COMP      2:STAT
3:TABLE
```

Figure A3.4 The fx-83's Mode menu will allow you to change the calculator's mode.

 (COMP)

Figure A3.5 Key entry for setting the calculator to COMP mode.

Note: When you first power up the fx-83 calculator it will be in 'MathIO' mode. This may be awkward for use with many basic engineering calculations. To change the mode to the more conventional 'LineIO' mode, you need to press the SHIFT button followed by MODE and 2.

Table A3.1 The display symbols used on the fx-83.

Display	Meaning
S	The keypad has been shifted by pressing the SHIFT key. The keypad will unshift and this indicator will disappear when you press a key.
A	The alpha input mode has been entered by pressing the ALPHA key. The alpha input mode will be cancelled and this indicator will disappear when you press a key.
M	This indicates that a value has been stored in the calculator's memory.
STO	The calculator is waiting for the input of a variable name. The calculator will then assign a value to this variable. The indication appears after you press SHIFT RCL STO.
RCL	The calculator is waiting for the input of a variable name. The calculator will then recall the value of this variable. The indication appears after you press RCL.
STAT	The calculator is in the statistical mode.
D	The default unit for angles is degrees.
R	The default unit for angles is radians.
G	The default unit for angles is grads.
FIX	The calculator has been set to display a fixed number of decimal places.
SCI	The calculator has been set to display a number of significant digits.
Math	The calculator has been set to the Maths input/output mode.
▼▲	Calculation history is available and can be replayed (or there is more data above or below the display).
Disp	The display currently shows an intermediate result of a multi-statement calculation.

Configuring the calculator setup

The fx-83 Setup menu allows you to control the way in which calculations are performed as well as the way that expressions are entered and displayed. The menu will let you work with a fixed number of decimal places (FIX), or with a fixed number of significant digits (SCI). You can also choose to use mathematical notation (MthIO) or conventional (line-based) notation (LineIO). The Setup menu is shown in Figure A3.6 whilst the keystrokes required to set the calculator to LineIO mode are shown in Figure A3.7.

SHIFT   SETUP   2   (LineIO)

Figure A3.6 Setup menu.

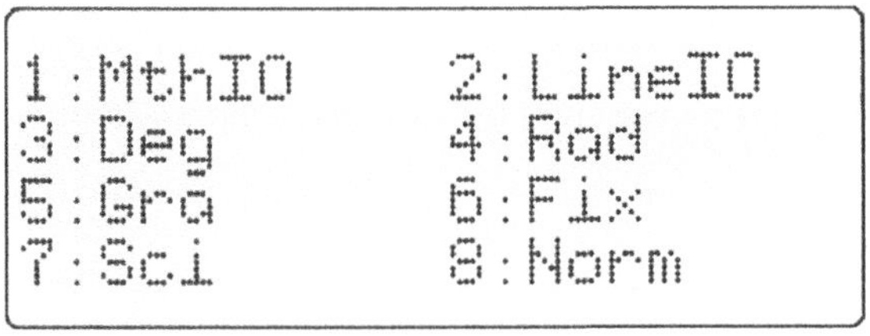

```
1:MthIO      2:LineIO
3:Deg        4:Rad
5:Gra        6:Fix
7:Sci        8:Norm
```

Figure A3.7 The keystrokes required to set the calculator to LineIO mode

Appendix 4
Useful websites

Chartered Institute of Plumbing and Heating Engineering (CIPHE)
www.ciphe.org.uk

Chartered Institution of Building Services Engineers (CIBSE)
www.cibse.org

Department for Business, Energy and Industrial Strategy (DBEIS)
www.gov.uk/government/organisations/department-for-business-
energy-and-industrial-strategy

EMTA Awards Ltd (EAL)
www.eal.org.uk

Engineering Council
www.engc.org.uk

Engineering Employers' Federation (EEF)
www.eef.org.uk

Health and Safety Executive (HSE)
www.hse.gov.uk

Institute of Cast Metals Engineers (ICME)
www.icme.org.uk

Institute of Marine Engineering, Science and Technology (IMarEST)
www.imarest.org

Institution of Engineering and Technology (IET)
www.theiet.org

Institution of Mechanical Engineers (IMechE)
www.imeche.org

Institution of Occupational Safety and Health (IOSH)
www.iosh.co.uk

Investors in People (IIP)
www.investorsinpeople.com

Science, Engineering, Manufacturing and Technologies Alliance (SEMTA)
www.semta.org.uk

Science Technology Engineering and Mathematics (STEM)
www.stemnet.org.uk

Society of Operations Engineers (SOE)
www.soe.org.uk

Index

Note: 'F' after a page number indicates a figure; 't' indicates a table.

Printed and bound by CPI Group (UK) Ltd, Croydon, CR0 4YY

22/10/2024

01777614-0004